Manufacturing Outsourcing

Asbjørn Rolstadås · Bjønar Henriksen
David O'Sullivan

Manufacturing Outsourcing

A Knowledge Perspective

 Springer

Asbjørn Rolstadås
Department of Production and Quality
Norwegian University of Science
  and Technology
S. P. Andersens vei 5
7491 Trondheim
Norway

Bjørnar Henriksen,
SINTEF Technology and Society
7465 Trondheim
Norway

David O'Sullivan
School of Engineering and Informatics
National University of Ireland
Railway Cottage Moycullen Co.
Galway 3017
Ireland

ISBN 978-1-4471-6108-0   ISBN 978-1-4471-2954-7 (eBook)
DOI 10.1007/978-1-4471-2954-7
Springer London Heidelberg New York Dordrecht

Printed on acid-free paper

Springer is part of Springer Science+Business Media (www.springer.com)

Preface

In many industrialized countries we have witnessed over recent decades the long-term decline in the percentage of employment in the manufacturing sector. We increasingly talk in terms of the "post industrialized" epoch. However, manufacturing in many respects is as important in industrialized nations as it has ever been. The need for products and services is increasing not decreasing. What is also happening is that due to increased efficiencies and automation, fewer people are producing more, than ever before. In addition of course new developing countries such as China, Brazil, and India are beginning to compete over a much wider range of competencies in manufacturing processes. China for example is no longer competing solely in the area of inexpensive products on the basis of low labour costs. Developing countries are now also competing for the design and manufacture of high quality products. The changes we are witnessing in the business environment, in society and life in general are making manufacturing more complex and are changing the basis of competition. Consequently, manufacturing is beginning to receive greater attention among politicians, businessmen, researchers, and the public in general in what Skinner once called a 'competitive weapon' for the growth and development of the economy through jobs creation and growth of economic output.

The basis for competition among industrialized economies for the continued growth of the manufacturing sector is changing and is shifting toward a strong focus on knowledge and in particular knowledge creation. Governments and politicians at different levels are beginning to engage in the enablement of innovations and continuous learning through educational systems, and skills development but also through various incentives to individuals and companies. We believe, however, that the manufacturing companies themselves have a key role to play and have the main responsibility for continuous innovation and knowledge creation that can make them more competitive in the global market place. One important reason for the focus on knowledge is that companies are becoming increasingly global in their outlook and how they manage their operations and that with new information and communications technology national borders are not the same barriers as they were before. Knowledge and knowledge creation can and will become the new 'competitive weapon' for industrialized countries.

This book focuses on manufacturing companies and how they can deal with an increasingly complex and changing business environment. It takes a fresh view of manufacturing from a knowledge management perspective that is both imaginative and practical in its potential to create competitiveness for most manufacturing organizations. The first two parts of the book present global trends and issues that together represent opportunities and treats for manufacturing organizations. We show that the traditional governance struggles to deal with many of the challenges that are emerging. One example of a major threat is the global financial crisis and yet another is the global environmental crisis. Individual companies need to consider strategies that reduce risks and vulnerability associated with these issues and the dramatic changes they imply. Companies can also develop strategies that open up windows of opportunities, for example related to sustainable manufacturing. What is important here is that manufacturing companies are able to develop strategies that accurately capture the manufacturing context and then develop capabilities that make the company a winner in the continuously changing market place.

Developing capabilities and positioning of the manufacturing company are based on a focus on knowledge, and ability to innovate in key areas. When we talk about capabilities we are increasingly talking about people, people that carry knowledge, create knowledge and translate knowledge into innovation. This knowledge is to a large extent 'tacit' and we often need to find ways of transferring the knowledge and making it more explicit. We see that different kinds of knowledge, knowledge from different sources and contexts, often need to be merged to create basis for innovation and more continuous improvement in manufacturing. We discuss how knowledge is one of the main strategic issues for manufacturing companies. When companies make major strategic decisions such as outsourcing or off shoring, knowledge aspects must be dealt with in a structured and strategic way. Outsourcing decisions are extremely important for a company, the pros and cons often have a complex interaction. Adding the knowledge dimension to an outsourcing decision for example, can have the effect of transforming the decision. This book will enlighten some key aspects related to knowledge and its importance in decision making.

The book covers both general and global trends in manufacturing and key aspects of knowledge from the strategic down to the shop floor level. When necessary, we have used aggregated statistics for knowledge in countries and regions for topics such as competitiveness, innovation, education, and environment. But we also want to bring the knowledge dimension down to a more practical level and present examples, frameworks, methods, and templates that can be applied by managers in their own environments. The last part of the book presents a number of 'cases' that show how manufacturing companies in different industries and contexts face different challenges when it comes to knowledge. These 'cases' are based on real companies, real situations, and real strategies.

The book is divided into five parts. Although the parts have a focus on knowledge, they aim to emphasize different facets of knowledge in manufacturing:

Part I: This part gives an overview of the book, our approach and the key issues that we discuss. The different chapters within this part describe the trends in manufacturing and forces we have to deal with in developing manufacturing strategies. This part also introduces key issues and terms that will be further investigated in the book including sustainability, paradigms, types of knowledge, and so on.

Part II: In this part we also present statistics about how companies and regions are doing when it comes to competitiveness, education, environment, and a number of other indicators. Important government initiatives and projects are presented that can be of interest to a number of policy makers in different industrialized nations. Important research fields that set the direction for innovations and knowledge creation financed by governmental bodies are also presented.

Part III: Developing manufacturing strategy can be a difficult and complex activity. It entails continuous decision making, which will ultimately lead to implementation and positive change. In this part of the book strategic decisions are discussed that focus in particular on outsourcing decisions. The knowledge dimension of the strategic decisions on structure and infrastructure are outlined.

Part IV: Innovation and knowledge are closely related and in this part we investigate both innovation and innovation theories and knowledge, and especially the link between innovation and knowledge. This part also presents several methods and approaches to deal with innovation and knowledge creation. The challenges we meet in strategies attached to different manufacturing paradigms are also discussed.

Part V: The final part of the book, presents chapters that show how the different aspects of manufacturing, outsourcing, and knowledge can have strategic, but also practical implications for companies. A number of 'cases' based on real companies show how they have to deal with different challenges depending on their strategic contexts, strategic decisions, and so on.

Contents

Part I Trends in Manufacturing

**1 Taking Control of the Company's Destiny
Through Knowledge** 3
   1.1 We are Sure about Some Things 3
   1.2 We Expect Some Things 4
   1.3 How Do We Deal with an Uncertain Future?. 5
   1.4 Paradigms and Capabilities 6
   1.5 Developing Capabilities Through Knowledge. 7
   1.6 The Knowledge Dimension of Manufacturing
       Outsourcing 8

2 Drivers for Change in Manufacturing 11
   2.1 Globalization, Extended Enterprise,
       Digital Business, Innovation 11
   2.2 New Versus Old Economy. 13
   2.3 Sustainability: On Top of the Manufacturing Agenda 15
   2.4 Quality and Productivity 15
   References 16

3 Manufacturing in a Strategic Context 19
   3.1 What is Manufacturing Strategy 19
   3.2 Generic Versus Specific Strategies 22
   3.3 Developing Capabilities. 24
   References 25

4 Evolving Paradigms in Manufacturing 27
   4.1 From Craft to Adaptive Manufacturing 27
   4.2 The Lean Paradigm. 29
   4.3 Sustainable Manufacturing. 30
   References 33

5   The Knowledge Dimension 35
    5.1   Capabilities Developed Through Knowledge 35
    5.2   Knowledge, the Basis for Innovations and Improvements 36
    References 39

Part II   The Engine Driving Industrial Change

6   Industrial Outlook 43
    6.1   Manufacturing in Turbulent Periods 43
    6.2   Technology Outlook 44
    6.3   Industrial Production: Who Takes the Lead? 45
    6.4   What Defines the Winners? 46
    References 51

7   Indicators and Initiatives for Industrial Renewal 53
    7.1   Education for Growth and Industrial Renewal 53
    7.2   The Lisbon Agenda 55
    7.3   Innovation Driving Industrial Change 57
    7.4   The Learning Economy 58
    7.5   Social and Environmental Renewal. 59
    References 61

8   Research Roadmaps in Manufacturing 63
    8.1   Research Resources in Manufacturing. 63
    8.2   Shift in the Manufacturing Research Agenda 64
    8.3   The IMS2020 Roadmap 65
    References 68

9   How Well Are We Doing? 69
    9.1   Measuring Performance in Companies 69
    9.2   Measuring Performance According to Strategy. 72
    9.3   Frameworks for Strategic Performance Measurement 74
    9.4   Decomposition, the Logic Behind the Strategy. 77
    9.5   Performance Measurement Implementation 78
    9.6   Performance in Supply Chains and Among Partners 82
    References 83

Part III   Outsourcing: Strategic Opportunities

10   Manufacturing Strategies, Created Through Decisions 87
    10.1   Structure and Infrastructure, Hard and Soft Elements
         of the Strategy 87

10.2 Structure, The Physical Manifestation of a Strategy 88
10.3 The Structural Prerequisites for How to Deal
with Knowledge. 89
10.4 Infrastructure, Exploiting the Structure 91
10.5 Structure and Infrastructure, a Complex Interplay
of Decisions. 92
References 93

11 Make or Buy? 95
11.1 What is Outsourcing? 95
11.2 Why Outsource 97
11.3 Off-Shoring 98
11.4 The Decision Process 99
References 100

12 The Geographical Footprint 101
12.1 Location of Facilities 101
12.2 Approaches to Location and Outsourcing Decisions 102
12.3 Focus on Process or Product 103
12.4 Where to Do What 104
References 107

13 Approaching the Partner Selection and Location Decision 109
13.1 Manufacturing Paradigm as Premise
for Outsourcing Decisions 109
13.2 Location Decisions and Partner Selection 110
13.3 Strategic Criteria in Location and Outsourcing Decisions:
An Example. 113
13.4 Location Criteria and Sustainable Manufacturing 116
References 116

14 Dealing with Complexity: Infrastructure Decisions 119
14.1 Outsourcing Need to be Accompanied
by Coordination Mechanisms. 119
14.2 Agency Theory 120
14.3 Mechanisms for Coordination 120
14.4 Coordination of Innovations and Knowledge Creation. 121
14.5 Implementing Outsourcing: Global Projects 123
References 125

Part IV   Innovation and Knowledge Transfer

15 The Innovation Process.................................. 129
15.1   What is an Innovation................................ 129
15.2   Innovation Processes................................. 132
15.3   Innovation and Centralized Versus Decentralized
       Knowledge Creation 134
15.4   Coordinating Innovation Processes: Project Models 134
15.5   Managing Innovation 136
15.6   Innovation and R&D Models.......................... 139
15.7   PDCA in Incremental and Radical Innovation 141
15.8   Manufacturing Paradigms and Innovation 143
References 144

16 What is Knowledge?................................... 145
16.1   Rationalism and Empiricism 145
16.2   Pragmatic and Combined Views on Knowledge........... 146
16.3   Knowledge and Information........................... 146
16.4   Terms Related to Knowledge.......................... 147
16.5   Basic Dimensions of Knowledge 148
16.6   Tacit and Explicit Knowledge 149
16.7   Knowledge Bases and Innovation...................... 150
References 153

17 Knowledge Creation.................................. 157
17.1   Putting Bits and Pieces Together 157
17.2   The Trade-Off Barrier................................ 158
17.3   Conflicts of Interests................................ 159
17.4   Knowledge Conversion 159
17.5   How to Integrate Knowledge Creation into
       the Development Model 160
17.6   Process Improvement and Knowledge................... 164
References 165

18 Knowledge Transfer and Distance...................... 167
18.1   What is Distance 167
18.2   The Cultural Challenge............................... 168
18.3   The Spatial Challenge of the Portfolio of Innovations....... 168
18.4   ICT Reduces Distance................................ 170
References 171

19 Knowledge Transfer and Manufacturing................. 173
19.1   A Model 173
19.2   Paradigms and Knowledge............................ 175

19.3   The Knowledge Dimension of Sustainable Manufacturing   177
References   180

20   Outsourcing and Sustainability   181
20.1   Energy Consumption on the Strategic Agenda   181
20.2   Energy Consumption: Process Technology   182
20.3   Outsourcing and Supply Chain Aspects   186
20.4   Improvement in Energy Consumption   188
20.5   Organizational: and Cross Organizational Learning   190
20.6   Sustainability at the Strategy Agenda   191
References   192

Part V   Cases

**21   Supply Chain Integration and Knowledge Transfer—A Case
from the Automotive Industry (Case 1)**   197
21.1   The Strategic Context   197
21.2   Business Systems and Types of Supply Chain Relations   199
21.3   Integration in the Truck Supply Chain   200
21.4   Fact Based Knowledge Transfer in the Supply Chain   202
References   202

**22   Quality Improvement in Craft Manufacturing—A Case
from Leisure Boat Manufacturing (Case 2)**   205
22.1   The Strategic Context   205
22.2   Quality Improvement as an Integrated Part
of Craft Manufacturing   207
22.3   Integrating Suppliers into Quality Improvement   208
22.4   Infrastructure for Knowledge Transfer   209
References   210

**23   Adaptive Manufacturing and Real Time Knowledge—A Case
from Furniture Manufacturing (Case 3)**   211
23.1   The Strategic Context   211
23.2   The Three Elements of Adaptive Manufacturing   213
23.3   Automatic Data Collection for Creating Knowledge   215
References   216

**24   Sustainable Manufacturing in SMEs—A Case from
Sportswear Manufacturing (Case 4)**   217
24.1   The Strategic Context   217
24.2   Competitive and Stable Workforce Through Inclusion   219

24.3   Manufacturing Units in China Included
         in the Sustainability Strategy 220
24.4   The Knowledge Dimension of the Sustainability Strategy 222

About the Authors 225

Index 227

Abbreviations

AHP	Analytical Hierarchy Process
APT	Automatically Programmed Tool
BEEM	Business Effect Evaluation Methodology
BRIC	Brazil, Russia, India, China
CAD	Computer-Aided Design
CAM	Computer-Aided Manufacturing
CEO	Chief Executive Officer
CIM	Computer Integrated Manufacturing
CIS	Commonwealth of Independent States
CNC	Computer Numerical Control
CSR	Corporate Social Responsibility
EMS	Energy Management Software
ERP	Enterprise Resource Planning
EU	European Union
GDP	Gross Domestic Product
GEM	Global Education in Manufacturing
GIS	Geographical Information Systems
GHG	Greenhouse Gas
GM	General Motors
HQ	Headquarters
ICT	Information and Communication Technology
IMF	International Monetary Fund
IMS	Intelligent Manufacturing System
IPR	Intellectual Property Rights
ISO	International Organization for Standardization
IW	Inclusive Work
JIT	Just in Time
LCA	Life Cycle Assessment
LCIA	Life Cycle Impact Assessment
MAPI	Manufacturers Alliance
MDP	Markov Decision Process

ML	Machine Learning
MNC	Multinational Corporation
NEST	Nature, Economy, Society, Technology
NUMMI	New United Motor Manufacturing
OECD	Organization for Economic Co-operation and development
OEM	Original Equipment Manufacturer
OM	Operations Management
PDCA	Plan-Do-Check-Act
PMI	Project Management Institute
PMO	Project Management Office
PPO	Production Paradigms Ontology
QFD	Quality Function Deployment
R&D	Research and Development
SLCA	Simplified LCA
SMED	Single Minute Exchange of Dies
Toe	Tonne oil equivalent
TPS	Toyota Production Systems
UNESCO	United Nations Educational, Scientific and Cultural Organization

Part I
Trends in Manufacturing

Manufacturing is going through an intense period of change and successful companies need to develop strategies for coping with this change. Knowledge has become the key driving force for change within manufacturing. It is through knowledge that we are able to develop capabilities and capture the opportunities in the business environment. This part of the book describes major trends in the use of knowledge, and why manufacturing is entering into a stage where knowledge aspects should be emphasized in strategic decision-making. Manufacturing outsourcing in particular stands to benefit from the deployment of knowledge for strategic developments.

Chapter 1
Taking Control of the Company's Destiny Through Knowledge

Abstract The context of manufacturing is increasingly complex and manufacturing has to deal with a continuously changing environment. There are some important aspects in manufacturing we see clearly today, and trends that we are quite sure about such as, globalization, extended enterprise, digital business and innovation, social responsibility and environmental issues. The challenge is to predict these developments and trends in more detail, making them relevant for industries and companies. Developing strategies is a way to deal with the future, the things we believe will happen, what we are not certain about but could expect, but also to have a kind of preparedness and flexibility for the unexpected. Manufacturing strategies are important for developing capabilities aiming to cope with a business environment characterized by change and increased complexity. Knowledge enables us to develop capabilities and to exploit the opportunities.

1.1 We are Sure about Some Things

The context of manufacturing is increasingly complex and manufacturing has to deal with a continuously changing environment. There are some important changes in manufacturing that we can see clearly today such as globalization, extended enterprise, digital business and innovation.

We are quite sure about a continuous technology developments within all fields of manufacturing. These developments in areas such as sensor technologies and wireless communications will result in new and improved products and processes that will change customer- and market perceptions and in turn lead to new requirements that can create a range of new opportunities for manufacturing companies. Technology will continue to progress in fields such as materials science, microelectronics, information technology, biotechnology and nanotechnology enabling manufacturers to innovate and offer better products and services to their customers in the future. These new technologies are to a large extent

A. Rolstadås et al., *Manufacturing Outsourcing*,
DOI: 10.1007/978-1-4471-2954-7_1, © Springer-Verlag London 2012

known and are all based on 'explicit' knowledge. When radically new products are to be manufactured we will have to create new processes that meet product requirements cost effectively. New technologies, as they emerge will have direct impact on emerging production processes. We see how ICT (Information and Communication Technology) has improved not only production capacity through automation but also enabled more intelligent and agile production systems through better planning and shop floor control. Customer choice is a powerful driver towards increased variety, perfect quality, instant off the shelf availability, and low cost. Manufacturing has to continuously respond to this desire through the use of 'agile' techniques.

Meeting customer demands is just one challenge facing the process of manufacturing systems design. Social responsibility and environmental issues are new issues that are becoming important for manufacturing. In this respect we will continue to see more regulations and incentives for more environmental friendly products and processes. These regulations, industrial standards and requirements will represent constraints but can also represent real opportunities for manufacturing companies that can adapt best to meeting them in the future.

There are also other market trends that will affect the manufacturing systems design process, for example, related to demography. New customer can now come from any part of the globe and societal needs within these new regions for most manufacturers will put pressure on industry to come up with fresh approaches to new products and services. Some of the new markets are a consequence of industrialization or growth of new regions. New market and societal needs also refer to areas such as safety and security, health care and health technology, energy supply and transportation.

1.2 We Expect Some Things

We are sure that in the future manufacturing need to be more innovative, deal with new technologies and more complex customer requirements. The challenge is to predict these developments and trends in more detail, making them relevant for industries and companies. For example, the environmental burden of processes and products is likely to be the subject of increased public scrutiny in the future. Manufacturing has to comply with stricter environmental regulation and might be confronted with new policy incentives to improve environmental performance. However, the concrete requirements (and opportunities) will be a result of political processes at national, regional and global levels, and could be difficult to predict. The public and market pressure for more environmental and socially responsible manufacturing will be a part of this picture. Changes in consumer preferences will demand the design of more sustainable products and services.

Another example is related to new technologies. There are many examples of new technologies but to what extent can manufacturing companies adopt these technologies and find application areas or markets for these technologies could be difficult to predict. Will existing customers' accept new technologies, or maybe they will be challenged due to a lack of skills, necessary infrastructure or ethical issues?

1.3 How Do We Deal with an Uncertain Future?

Developing strategies is a way to plan for the future, the things we believe will come, what we are not certain about but could expect, but also to have a kind of preparedness and flexibility for the unexpected.

Manufacturing strategies have to capture many uncertainty issues for manufacturers: controlling risks and quality in outsourcing and offshoring, supply chain management, prices of raw materials and especially energy. Through the manufacturing strategies the companies have to find next generation efficiencies and perfection in process management. The future of engineering and robotics, 3D modeling, faster prototyping, speed to market and innovation are example of innovative strategic priority areas. Manufacturing strategy also has to be a guide and enabler for protecting the environment in a profitable way. Depending on industrial sector or region there are many other developments and trends that will be used in the development of manufacturing strategies. It is difficult to get a good picture of the future of manufacturing industries, especially since disruptive and unexpected events and incidents will occur.

One way to try to construct a picture of the future could be through scenarios and foresight processes. Foresights have been popular in many industries. Figure 1.1 illustrates a foresights 'compass' for the Norwegian leisure boat industry. Developing the foresight was an extensive process involving key players in the leisure boat industry as well as researchers and experts from within and outside the industry. The most important outcome of the foresight process was probably an increased knowledge and awareness for the participants about trends, developments and possible futures. The compass shows how strategies could be made from the positioning trends according to the following dimensions:

- The global economic situation
- To extent to which companies are proactive and innovative.

The first dimension is based on a common understanding of strong correlations between market demands and the general economic conditions. Companies and the industry as a whole are not in a position to influence these conditions—they are referred to a 'contextual constraints'. The second dimension is very important in policy and strategy making for the industry and companies since it indicates whether the companies' tries to influence their future through innovations and proactive actions—referred to as 'transactional constraints'. Each of the four segments in the compass scenarios were developed and relevant strategic issues and measures where described to create strategic awareness. The scenarios ranged from a picture of an innovative, high tech, competitive and growing industry called

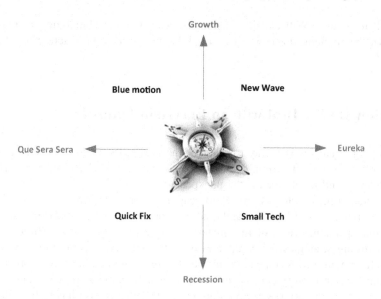

Fig. 1.1 Foresight for the Norwegian leisure boat industry

'New Wave', to a much more defensive and reactive industry in the 'Quick Fix' scenario and observed particularly in times of global recession.

We know that there are alternative ways of including the different, trends and developments, opportunities and treats in manufacturing strategies. A company could adjust strategies and follow the roadmap developed by other companies. An alternative approach is to be proactive and try to influence or change some of the business environment. This could be challenging and requires significant resources, but at least companies could try to benefit from what we see in trends and developments around us. This approach is closely related to the resource-versus market view on manufacturing strategies, where the resource oriented approach means that the company tries to influence customers and business environment to better fit the companies' capabilities. At the same time the resources and capabilities are further developed to fit the markets. The market view approach alone has more of its focus on adjustments to the business environment.

1.4 Paradigms and Capabilities

A manufacturing strategy can be seen as set of decisions relating to a company's structure and infrastructure. In this respect, structure concerns the vertical integration of operations, facilities and locations, capacity, and process technology, while infrastructure represents organizational and human aspects. Manufacturing strategies are important for developing capabilities aiming to cope with a business environment characterized by change and increased complexity.

In a strategic context, manufacturing is influenced by internal and external factors. The way manufacturing deals with these factors will often be reflected in paradigms. Paradigms represent principles and approaches which are not only related to practical issues but also to a more fundamental way of approaching strategic challenges. Craft manufacturing, mass manufacturing, lean manufacturing, adaptive manufacturing and sustainable manufacturing, are currently common paradigms in manufacturing. Even paradigms that have seen more popularity in the past such as craft manufacturing and mass manufacturing have particular facets which can be deemed important in current approaches to modern manufacturing.

To gain a good position in the market companies need to develop core capabilities. As we shall see from cases later in this book the relevant manufacturing capabilities depends on the strategic context. Principles and the ideas of 'how things should be done' and reflected in various paradigms are premises for what is considered as the important capabilities developed through manufacturing strategies.

1.5 Developing Capabilities Through Knowledge

Knowledge is important when identifying and developing capabilities and to exploit the opportunities companies face in their business environment. To create and transfer knowledge about what is happening at the market place, stakeholders' expectations, and so on, it is important to establish a good strategy and to continuously adjust and improve it. This means for example to anticipate and foresee customers' preferences, governmental regulations, and discover technological innovations. Among other things, this is knowledge that is important in order to get a picture of what capabilities we need to develop. However, maybe even more important is the knowledge creation within the company, supply chain and other stakeholders aiming to actually develop and continuously improve capabilities. The knowledge dimensions of strategic decisions are not only reflected in day-to-day operations and knowledge creation for continuous improvement, and also in research and development projects.

Strategies are about taking decisions aimed at developing capabilities. Decisions on the structure and infrastructure of manufacturing are prerequisites for what kind of knowledge should be focused on, but also strategies around the knowledge creation process itself:

- The kind of knowledge to be focused on within manufacturing is a consequence of process technology. Process technology is in turn a consequence of stakeholders' requirements and demands

- Decisions on facilities, especially location and vertical integration are infra-structure decisions that define who will be key actors in knowledge creation and consequently the challenges related to knowledge transfer
- Infrastructure decisions are to a large extent about organizing and coordinating activities in a way that enables knowledge transfer between the main actors in a manufacturing process.

Knowledge transfer is important for a company aiming for flexibility and capabilities for capturing the opportunities in their business environment. The knowledge dimension of the manufacturing strategy is important, but the knowl-edge challenges will often differ between strategies. This also means that the enablers for knowledge transfer will be different and should be an integrated part of the manufacturing strategy.

The knowledge dimension in manufacturing is largely a question of how to deal with the transfer challenges of tacit and explicit knowledge. Manufacturing par-adigms have different challenges related to these issues. For example, craft manufacturing and to some extent lean manufacturing will base more of their improvement and development on tacit knowledge than we see in strategies based on mass manufacturing and to some extent sustainable manufacturing. As the manufacturing strategies are increasingly focused on capabilities related to flexi-bility, customizations, innovations, and other less tangible aspects, the measure-ment of knowledge will present new challenges.

1.6 The Knowledge Dimension of Manufacturing Outsourcing

Make or buy decisions are one of the most important strategic decisions for a company. The pros and cons of outsourcing have been discussed in the literature for a long time. What has been less discussed is the knowledge dimension of manufacturing outsourcing. However, researchers have argued for increasing the importance of knowledge aspects of outsourcing and its strategic competitive importance in the market place. This means that the company needs to stimulate knowledge creation but also to have control of knowledge creation processes that are particularly important for developing manufacturing capabilities. Knowledge transfer is difficult within one organization but obviously even more difficult when important operations are outsourced to locations maybe far from corporate office and other facilities.

Flexibility and customization are increasingly important in manufacturing and in strategies aiming to develop such capabilities. Decentralization and employee involvement will often be determinants for success. So how do we design and motivate for such knowledge creation in a strategy based on manufacturing out-sourcing? First of all the knowledge dimension should be treated seriously in manufacturing outsourcing decisions. Secondly, enablers for knowledge transfer need to be integrated within the strategic decisions.

The knowledge transfer challenges will be related to the tacit–explicit dimension. In a decentralized structure, where the knowledge creation process is incremental, and involves people at shop floor level, the knowledge will often be tacit. In this situation there is a need to extend cycles of learning from the individual or team level to the organizational level. Normally, the transfer of knowledge will require some kind of codification of the knowledge, making it more explicit. The transfer of explicit knowledge is enabled by integrated systems. However, the adaption of this knowledge in the concrete working environment in another organizational unit may be difficult and represent a particular challenge for tacit knowledge. The knowledge must be transferred in a way that reflects the context where it is supposed to be adapted. Job rotation and 'knowledge brokers' are ways of making it easier to understand the units and people that are supposed to receive and use knowledge. Centers of excellence are other arenas where explicit knowledge related to new processes can be explained and transferred to the working environment of people from different manufacturing units.

Knowledge creation is essential to survive in the market place and to capture the opportunities in the future. Traditionally this has meant finding ways of capturing as much explicit knowledge as possible through ICT based reporting and management systems. With increased focus on flexibility, customization, and so on, the use of tacit knowledge becomes increasingly important, but is extremely challenging when outsourcing processes to other organizations. Even if enablers for knowledge transfer between different organizations exist, and we try to get control of knowledge creation in key areas, we must accept that some knowledge transfer must be based on personal relationships within and between organizational units.

Chapter 2
Drivers for Change in Manufacturing

Abstract During the last two decades the markets for industrial goods have become saturated with competing alternatives. Quality and price are still important but customers' needs as well as other stakeholders' requirements have changed. This has led to a new and dramatically changed competitive situation. Flexibility and the ability to capture opportunities are important issues in the continuously changing and complex business environment in the new economy, and are often considered as basic elements of today's manufacturing strategies. Companies need to constantly define and redefine their position in the supply chain since changes and innovations are evolving faster and faster. Increased competition and complex market demands also complicate this environment. Companies are constantly considering which partners are the right ones to provide value for their stakeholders and this is demanding flexibility and agility to form and reform partnerships along the supply chain. In outsourcing and partner selection new criteria will need to be added such as the ability to create, transfer and act on new knowledge.

2.1 Globalization, Extended Enterprise, Digital Business, Innovation

Over the last two decades the markets for industrial goods have become saturated with new products and services. This could be said to have its origin after the Second World War, when there was a large demand for industrial goods. In this post war era mass manufacturing allowed products to be manufactured cheaply. This has changed over the decades and today there are arguably more products available than there is a need for. A high-quality product at a reasonable price will no longer automatically find a market. Customers' expectations have changed. Of course this has led to a new and dramatically changed competitive situation. Productivity and competitive advantage have become a major issue. The new competitive situation has forced the development of new business operation strategies. At an Intelligent Manufacturing System (IMS) meeting in Zurich in 2007, leading academics and industrialists from around the world identified four major drivers for change [1]:

- *Globalization* of manufacturing including decisions on which manufacturing plants should go offshore and which should remain onshore.
- *Extended enterprises* and the way manufacturing organizations are increasingly becoming collaborators in a value chain.
- *Digital business* and its enormous effect on change and potential for even more change.
- *Innovation* with its ability to not only increase productivity but also enable innovative new processes for new types of products.

These forces reinforce each other and have increased the competitive pressure to improve and change and represent a business environment wherein companies have to create and implement their strategies. If we go beyond these forces, we see that increased availability and capacity to analyze and make use of data and information has been a common denominator for the changes in competitive situation and strategies.

Globalization includes two aspects: markets and manufacturing. Markets have been global for a while. Lately, manufacturing has also become global. Enterprises are moving in a multinational direction. Better information and communication technologies (ICT), and access to global markets have encouraged companies to rely on distributed networks and supply chains that are not only long, but also shared among different firms, and in which not necessarily anyone owns the whole chain [2]. The basic supply chain is rapidly evolving into what is known as a supply chain network [3]. It is especially true in a situation where flexibility is required that such networks might become crucial parts of a manufacturing strategy. Such types of networks rely heavily on knowledge sharing and represent important coordination challenges.

Extended enterprises are spread all over the world. They are located where the total conditions for operation are most favorable. The same is true for design, product development, technology development, and other crucial business processes in the enterprise. However, most favorable location for different units of the extended enterprise, depends not only on the products, customer needs, and market size, but also on the production process, factor inputs, and increasingly also a business environment that can stimulate innovation. In a virtual enterprise the modern enterprise will see its customers and suppliers as part of their own company by establishing strategic alliances. Kurihara discusses next generation manufacturing enterprises and sees support for the virtual enterprise as a key characteristic [4].

Customization is a market focus aimed at providing the customer with a product as close as possible to his point of demand as we have seen in craft manufacturing, but without losing the advantage of low cost mass production. This concept is further developed towards customer focus or customer satisfaction. The enterprise tries to gain new market share by surpassing the customer's expectations and constantly establishing new market standards. This means excellence in all respects including zero defects, short delivery times, customization and low costs, and where employees are trained to think customer satisfaction. The global situation with the virtual or extended enterprise, extends the customer satisfaction approach beyond the customer and includes stakeholders such as suppliers, vendors, after sales, financial institutions, and even local authorities. To have the best suppliers is considered an asset. There is therefore an open competition in the marketplace not only for customers, but also for suppliers and owners [5]. This customization and agility in supply networks that we see in virtual manufacturing obviously requires integrated and flexible ICT systems between the various stakeholders. The stakeholders involved in the production, distribution, after sales, and so on will need to have effective ways of sharing data and information. However, the more intangible aspects related to the customers or market segments also need to be captured. This knowledge could be more difficult to capture and share within a virtual enterprise since it is often less explicit.

We have seen that social responsibility and environmental issues have received increased importance not only in academic and political forums, but are now also represent the basic premises in how manufacturing systems operate. This is a result of more public attention on these issues and a more common understanding of issues such as limited resources, global warming and other effects of environmental imbalances. Corporate social responsibility and sustainable manufacturing are now more than just a way for companies to respond to new regulations and standards. We see that innovation in product and process development are increasingly focusing on sustainability, and consequently also set the direction for knowledge creation in manufacturing. In the future we will see that social responsibility and sustainability are even more than today will represent significant business opportunities for manufacturing companies. We will see more business models where companies build relationships with key stakeholders and customers aiming to make a true difference when it comes to social responsibility and sustainability.

2.2 New Versus Old Economy

Hayes et al. [2] have described the evolving "new economy," and the characteristics they identify that differentiate between the "new" versus "old" economy are summarized in Table 2.1.

Flexibility and the ability to capture opportunities are important issues in the continuously changing and complex business environment in the new economy, and are often considered as basic elements of modern manufacturing strategies. Flexibility can be in terms of products, order volumes, speed and responsiveness. In such cases a company must be able to offer a wide product range, deliver non-standard or customized products, and/or take the lead in introducing new products [2].

Table 2.1 Old economy versus new economy operations [2]

Issue	Old economy	New economy
Unit of analysis	An operating unit	A network of semi-independent players
Goal	Sell product/services	Develop ongoing relationship with customers, suppliers, and complementors
Domain of operational management (OM)	Product and processes	Systems of complementary products provided by different organizations in networks
Dominant operational management activity	Managing flows through a stable process	Managing the dynamics of highly flexible products through ever-changing processes and networks
Operational management tools	Flow analysis, scheduling, expediting, and so forth	Project management, negotiating, building consensus, designing incentives, and so forth
Primary measures of performance	Incremental unit cost and "quality" (i.e., low defects and/or high performance)	First unit cost and "acceptable quality" (i.e., low defects, ease of use, and improvability)
Competitive imperative	Achieve superiority along some valued dimension(s)	Get high volume quickly and induce others to support one's product/network
Performance improvement	Continuous improvement using PDCA cycles and other Kaizen tools	Learning across development projects
Competition	"Prevail" through differentiation	Jointly prosper through collaboration, resulting in a dominant standard

Larger companies may make other strategic choices through volume flexibility, exploiting an ability to accelerate or decelerate production very quickly and juggle orders in order to provide unusually rapid delivery. Mass customization based on module-based manufacturing can be a strategy to gain flexibility together with volume effects [6]. To meet increased complexity and customer pressure, Original Equipment Manufacturer (OEMs) and retail companies have handed more responsibilities to their suppliers in order to maximize the efficiency of their supply chains [7]. Suppliers have a derivative demand and must understand not only the driver who ultimately buys the end product, but also how the economics of their products increase the OEMs' profits. This requires not only knowledge transfer within a supplier's own company, but also a collaborative supply chain approach around where knowledge is.

From Table 2.1 we might think that in the future, outsourcing and location issues are not as important as they were in the past, since we in the future will face a situation where more of the activities are performed in collaboration between independent partners. But this is not necessarily the case. Companies need to constantly define and redefine their position in the supply chain since changes and innovations are occurring faster and faster. Increased competition and complex market demands are also substantiating this picture. The need for more flexibility requires that companies need to constantly consider whether their partners are the right ones. However, in outsourcing and partner selection, new criteria will be added such as the ability to create, transfer and act on new knowledge.

2.3 Sustainability: On Top of the Manufacturing Agenda

The over-exploitation of resources is one of the most important challenges facing us today. Increased consumption and more urban population raise environmental issues, for example issues related to water supply. Another area where change is observable is on the global climate, where the next 10 years are considered crucial for obtaining lasting emission reductions. The global energy supply consumption and supply will continue to be dominated by oil, gas and coal, but we are heading into a future decade marked by a transition towards low- carbon energy [8].

These global challenges also represent opportunities for innovation and improved technologies for recycling and better utilization of waste, energy efficiency, alternative materials, and so on. In sustainable manufacturing these challenges are basic elements of the manufacturing strategy.

Sustainability has definitely reached the strategic planning agenda in manufacturing [9]. To some extent the sustainability discussions have been focusing on quite radical changes in manufacturing and the "way of living". The availability of scarce resources, the need for a stable and educated workforce and a healthy environment are examples of issues that have often a direct impact on a company's productivity and the quality of products. We believe that sustainability will be an even more important premise for manufacturing in the future. This is not only because regulations and public pressure requires so, but also as a consequence of the increased direct impact of social and environmental aspects on the economic bottom line in the future.

2.4 Quality and Productivity

There are many ways companies can succeed in the market place. Garvin [10] has proposed eight dimensions or categories of quality that can serve as a guideline or framework to identify customers' requirements in manufacturing:

- *Performance*—is about the primary operating characteristics and to which extent the product performs to its standards
- *Features*—are the additional benefits through the product
- *Reliability*—is the probability of perform well and consistently over a specified period of time

- *Durability*—is about how well the product will last with daily use, the amount of use one gets from a product before it physically deteriorates or replacement preferable
- *Conformance*—describes the degree to which physical and performance characteristics of a product match any agreed internal or external specifications such as safety regulations and laws
- *Serviceability*—is the prerequisites for speed, competency, and competence of repair, if the product is easy to service and the organization offers enough service support
- *Aesthetics*—is a dimension that could be both tangible and intangible focusing on how a product looks, feels, sounds, tastes, or smells

Perceived quality—could be even more difficult to manage as it is based on subjective assessments resulting from image, advertising, brand names, etc. that could differ between people, regions, and change over time.

Making products meet customers' requirements doesn't help much if we are unable to produce them at costs accepted by sufficient number of customers. This in turn raises productivity issues, issues that have always been a basic premise for manufacturing. Productivity is basically about how well a company or an organization transforms input into output that meets with customers' requirements. Innovations and improvements in quality and productivity will continue to be an important aspect of manufacturing. And as the market place is continuously changing with increased competition, quality- a productivity improvement will be even more important.

When trends and drivers for change in manufacturing are described they are to a large extent describing how quality and productivity could be influenced by changes in the business environments, through innovations to products, processes, services and organization.

References

1. O'Sullivan D, Rolstadås A, Filos E (2011) Global education in manufacturing strategy. J Intell Manuf 22(5):663–674
2. Hayes RH, Pisano GP, Upton D, Wheelwright SC (2005) Operations, strategy, and technology—pursuing the competitive edge. Wiley, Hoboken
3. Kuglin FA, Rosenbaum BA (2000) The supply chain network @ internet speed: preparing your company for the internet revolution. AMACOM, NY
4. Kurihara T, Bunce P, Jordan J (1996) Next generation manufacturing systems in the IMS program. In: Okino N, Tamura H, and Fujii S (eds) Advances in production management systems, IFIP WG5.7, pp 17–22
5. Bredrup H (1995) Performance evaluation. In: Rolstadås A (ed) Performance management: a business process benchmarking approach. Chapman and Hall, London, pp 191–198
6. Zagnoli P, Pagono A (2001) Modularization, knowledge management and supply chain relations: the trajectory of a European commercial vehicle asseler. Actes du Gerpisa 32:45–64

7. Kennedy JV (2003) Broadening value added to include solutions, customized products, and related services. In: Duesterberg TJ, Preeg EH (eds) U.S. manufacturing. The engine for growth in a global economy. A project of the manufacturers Alliance/MAPI. Praege, Westport, Connecticut, pp 95–121
8. Det Norske Veritas (2010) Technology outlook 2020. DNV Research & Innovation, Høvik
9. Taisch M, Cassina J (2010) Action roadmap on key area 1, 2 and 3. In: IMS2020 project report, POLIMI, Italy
10. Garwin DA (1987) Competing on the eight dimensions of quality. In: Harvard business review, vol 65, issue 6, pp 101–109

References

Chapter 3
Manufacturing in a Strategic Context

Abstract To deal with the forces for change and to find roadmaps in a complex and changing competitive landscape companies need a vision, objectives and strategies across different fields of endeavour. One such field for almost all companies is related to 'who should do what' in relation to the procurement of the right products for customers. There is no uniform definition of manufacturing strategy and there are many different approaches. A key dimension is whether a given strategy has an external market view or a resource-based view. A combination of market based and resource based strategies are desirable where the company is able to capture market needs and achieve high volume production at the same time as they attempt to line up demand with the specific resources and capabilities of the company. Manufacturing strategies are essentially about finding ways to utilize resources in the most effective way and to develop flexible and agile capabilities. New paradigms can set directions for development, such as regarding 'what should be important' or 'what is the way to do things', but will normally also function as guides and principles for how to progress in a defined direction.

3.1 What is Manufacturing Strategy

To deal with the forces for change and to find development roadmaps in a complex and changing competitive landscape companies need a vision that can be clearly articulated using strategic objectives and performance indicators across a number of different areas. These strategies are not necessarily large written documents, but could be enabling statements and supports for informed decision making across critical areas. One such critical area for almost all companies is related to who should do what along the supply chain, what components or services to procure for the different categories of customer.

A. Rolstadås et al., *Manufacturing Outsourcing*,
DOI: 10.1007/978-1-4471-2954-7_3, © Springer-Verlag London 2012

Strategy as a research field has a long history and can be traced back to the strategy of war, and the ancient Greeks. Many approaches and techniques within strategy that have been developed for warfare are still in use today, such as for example game theory and scenario thinking. The history of manufacturing strategy is much shorter, but even if manufacturing strategy as a research field has been recent, manufacturing has always been part of the strategic context for companies. Skinner [1] has identified five decision areas for strategy development:

- Plant and equipment
- Production planning and control
- Labor and staffing
- Product design/engineering
- Organization and management

Leong and Ward [2] argue that a multifaceted view on manufacturing strategy is needed and describe the following perspectives:

- Planning
- Proactiveness
- Pattern of actions
- Portfolio of manufacturing capabilities
- Programs of improvement
- Performance measurement.

Even if a manufacturing strategy could focus on a certain issue, such as quality or supply chain, the definitions of manufacturing strategies today are wider than the traditional functional definition, wiping out barriers between manufacturing and say marketing, Research and Development (R&D), finance, and so on. Organizational boundaries between the stakeholders both inside companies and between partners along the supply chains are becoming less important when creating manufacturing strategies.

There is no uniform definition of manufacturing strategy and there are many different approaches to it. A main dimension is whether a given strategy has an external market view or a resource-based view. The differences between these two perspectives, as identified by Thun [3], are illustrated in Fig. 3.1. The external perspective has a focus on markets and customer requirements, and the roots can be seen in the structure–conduct–performance approach advocated by Mason [4] and Bain [5]. Porter [6], with his concept of "generic strategies," could be regarded as an advocate of the external view with his focus on "How competitive forces shape strategy", cluster theories, etc. Hill [7] describes how this external view of strategy has been dominating manufacturing in recent years, where sales and marketing define what has to be produced and to which requirements, and where manufacturing simply has to find ways of fulfilling these requirements.

Contrary to the *market-based* view, the *resource-based* view is an internally pursued perspective. Hayes et al. [8] argue for a resource-oriented perspective on strategy:

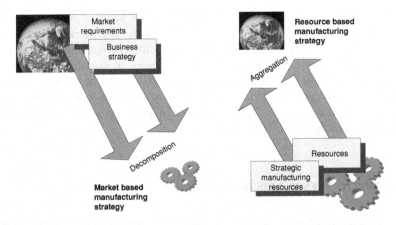

Fig. 3.1 Market based versus resource based manufacturing strategies

> The implication of a capabilities based approach is that competitive battles are won not in
> the boardroom but in the laboratories, on factory floors, at service counters, and in
> computer rooms. The role of the operations function expands from being simply the
> implementer of strategy to providing the foundation for—indeed, becoming a driver of—
> successful strategic attacks and defenses. [8, p. 56]

Based on data from an IMS survey, Cagliano et al. [9] concluded that the
'capability-based' manufacturing strategy was the one gaining popularity, in
contrast to the market-based strategy. Mintzberg and Quinn [10] argue for balancing internal and external perspectives in their definition of manufacturing
strategy. They define strategy as:

> [A] pattern or plan that integrates an organization's major goals, policies, and action
> sequences into a cohesive whole. A well-formulated strategy helps to marshal and allocate
> an organization's resources into a unique and viable posture based on its relative internal
> competencies and shortcomings, anticipated changes in the environment and contingent
> moves by intelligent opponents. [10, p. 3]

Evidently a combination of market based and resource based strategies are
desirable where the company is able to capture the market needs and get high
volumes at the same time as they are in line with the specific resources and
capabilities of the company. Many attempts and propositions have been made for
an integrated manufacturing strategy approach aiming for a strategic fit of market-
based aspects as well as resource-based issues. Hill [8] describes a stepwise
approach through their concept of "order qualifiers" and "order winners".

The definition and perspective on manufacturing strategy are reflected in their
approach to knowledge and the types of knowledge that they are focused on, such
as internal resources or markets, and who should be involved in knowledge creation. For example could outsourcing decisions in a market based strategy, focus
on finding partners located in promising geographical markets with the right
capacity and cost/quality performance. In a resource based view on manufacturing

strategy the outsourcing decisions would highlight the partners' ability and willingness to improve capabilities, and to identify and develop markets. In the first case the ability to create knowledge from the different markets will be very important for example related to global differences in customer preferences, necessary product adjustments, possible new markets, etc. In the resource oriented strategy the partners' contribution to innovations and knowledge creation is to develop capabilities through their own research facilities or relationships with research organizations and universities.

3.2 Generic Versus Specific Strategies

Michael Porter [6, 11] has described a category scheme consisting of three general types of strategies commonly used by companies. These strategies are defined along two dimensions: strategic scope and strategic strength or competitive advantage, with "cost leadership," "differentiation," and "focus strategies" targeting one or a few particular market segments.

Porter claims that combining multiple strategies is successful only when combining market segmentation and a product differentiation strategy. Combinations such as cost leadership with product differentiation are hard to implement due to the potential conflict between cost minimization and the additional cost of value-added differentiation. However, Hayes et al. [8] argue that such a mass customization is one possible outcome of modern flexible manufacturing principles. Throughout the 1980s and early 1990s the need for flexibility was increasingly emphasized, and Gerwin [12] presents four generic flexible strategies:

- *Adaption*—a defensive response to environmental uncertainty
- *Redefinition*—a proactive strategy aims to change the competitive situation by, for example, trying change customers' expectations
- *Banking*—a defensive/proactive strategy where investments, for example in alternative processes, are made to meet future needs for flexibility
- *Reduction*—a proactive strategy that is often seen in lean manufacturing when uncertainty is reduced through, for example, investments in long-term customer relationships.

Unlike Porter's generic strategies, specific strategies are not exclusive and more than one strategy is likely to be pursued at a time, as Kennedy [13] exemplifies:

- *Supply chain optimization*, involves outsourcing of activities to others that do them better
- *Adding a service component to the products*, aims to capture more of the total value received by the customer
- *Technology leader*, close to Porter's differentiation strategy
- *Integrating products into the information web*, creating new opportunities for products to capture, transmit, and act on information

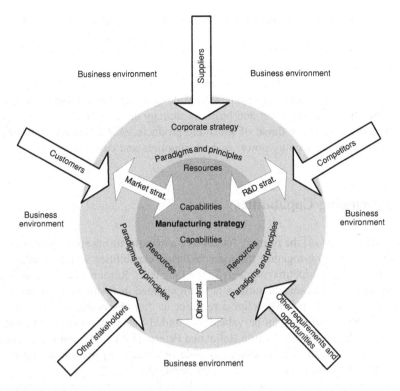

Fig. 3.2 Manufacturing in a strategic context

- *Customization*, most often targets customers who tend to pay a premium for products that appeal to their unique self-image.

Figure 3.2 shows the main elements that influence manufacturing strategies. The business environment is changing and increasingly complex in most industries. It represents both opportunities and threats, which are basic elements of the strategic context. Companies encounter these in their relations with customers, suppliers and other stakeholders but also through the strategies and activities of competitors.

Manufacturing strategies are basically about finding ways to use resources in the most effective way and to develop capabilities [14]. Paradigms set directions for development, such as regarding what should be important or what is the way to do things, but will normally also function as guides and principles for how to progress in a defined direction.

Manufacturing and manufacturing strategies are increasingly integrated in the other processes in a company. Such integration means that manufacturing is not only influenced by decisions within marketing, Research and Development (R&D) and Human Resources (HR), but also vice versa. Manufacturing decisions are often also premise providers for overall business strategies.

Figure 3.2 is not a complete model, but illustrates important elements shaping manufacturing strategy. There are several facets not emphasized in the figure, such as socio-cultural factors. How and to what extent the different elements shape a given manufacturing strategy are company specific [8, 12], and depend on the situation in industry concerned [6], its market position [15] and stakeholder [16, 17]. Porter's [6] five forces framework is designed for the analysis of business environments at industry level and business strategy development, and comprises the following forces: the threat of substitute products, established rivals and new entrants, and the bargaining power of both suppliers and customers.

3.3 Developing Capabilities

According to Tait [14] the primary function of a manufacturing strategy is to guide a business in putting together the manufacturing capabilities which enable it to pursue its chosen competitive strategy in the long term. Resources and capabilities are related terms, but several authors have highlighted distinctions between them.

Hoopes et al. [18, p. 890] define a resource as "an observable (but not necessarily tangible) asset that can be valued and traded—such as a brand, a patent, a parcel of land, or a licence," while Helfat and Peteraf [19, p. 999] define a resource as "an asset or input to production (tangible or intangible) that an organization owns, controls, or has access to on a semi-permanent basis".

Makadok [20] defines a resource as an observable asset that can be valued and traded, while a capability is not observable and cannot be valued. However, it is difficult to draw a clear distinction between resources and capabilities, and there is no uniform definition of capabilities. This is to some extent explained by capabilities being unique to each company, and that they are not easily recognized and categorized. Capabilities are difficult to imitate or transfer, rendering them valuable. Swink and Hegarty [21] have supported this view by defining capability as organizationally specific and internally developed. Capabilities derive less from specific technologies or manufacturing facilities and more from manufacturing infrastructure: people, management and information systems and learning [22].

Hayes and Upton [23] divide operating capabilities into the following types:

- *Process based* (e.g., the ability to provide advantages such as low costs and high quality)
- *Systems based* (e.g., the ability to create short lead times and to customize)
- *Organizational based* (e.g., the ability to master new technologies and introduce new products)

In contrast Helfat and Peteraf [19] classify capabilities as either "operational" or "dynamic", an operational capability generally involves performing an activity. This could for example be, manufacturing a particular product, while dynamic capabilities, as defined by Teece et al. [24], include activities such as to build, integrate, or reconfigure operational capabilities.

From the above it can be seen that capabilities are closely related to knowledge, and that resources could be regarded as capabilities if we have the knowledge to exploit them. We also understand that capabilities to a large extent are carried out by people and the knowledge, skills, ability and motivation they have for performing them.

References

1. Skinner W (1985) Manufacturing: the formidable competitive weapon. Wiley, New York
2. Leong KG, Ward PT (1995) The six Ps of manufacturing strategy. Int J Oper Prod Manage 15(12):32–45
3. Thun JH (2008) Empirical analysis of manufacturing strategy implementation. Int J Prod Econ 113(1):370–382
4. Mason ES (1939) Price and production policies of large-scale enterprise. Am Econ Rev 29(1):61–74
5. Bain JS (1959) Industrial organization. Wiley, New York
6. Porter ME (1980) Competitive strategy: techniques for analyzing industries and competitors. Free Press, New York
7. Hill T (2000) Operations management: strategic context and managerial analysis. Macmillan, London
8. Hayes RH, Pisano GP, Upton D, Wheelwright SC (2005) Operations, strategy, and technology—pursuing the competitive edge. Wiley, Hoboken
9. Cagliano R, Acur N, Boer H (2005) Patterns of change in manufacturing strategy configurations. Int J Oper Prod Manage 25(7):701–718
10. Mintzberg H, Quinn JB (1996) The strategy process: concepts, contexts, cases, 3rd edn. Prentice Hall, Upper Saddle River
11. Porter ME (1985) Competitive advantage creating and sustaining superior performance. Harvard Business School Press, New York
12. Gerwin D (1993) Manufacturing flexibility: a strategic perspective. Manage Sci 39(4):395–410
13. Kennedy MN (2003) Product development for the lean enterprise: why Toyota's system is four times more productive and how you can implement it. Oaklea Press, Richmond
14. Tait N (1997) Handling the sourcing decisions: lowest cost is not always the answer. Financial Times, Oct 15th, p. 13
15. Berry W, Hill T, Klompmaker J, McLaughlin C (1991) Linking strategy formulation in marketing and operations: empirical research. J Oper Manage 10(3):294–302
16. Bridges S, Marcum W, Harrison KJ (2003) The relation between employee perceptions of stakeholder balance and corporate financial performance. SAM Adv Manage J 68(2):50–55
17. Suntook F, Murphy JA (2008) The stakeholder balance sheet: profiting from really understanding your market. Wiley, Chicester
18. Hoopes D, Hadsen T, Walker G (2003) Guest editor's introduction to the special issue: why is there a resource-based view? Toward a theory of competitive heterogeneity. Strategic Manage J 24(10):890–904
19. Helfat CE, Peteraf MA (2003) The dynamic resource-based view: capability life cycles. Strateg Manag J 24(10):997–1010
20. Makadok R (2001) Toward a synthesis of the resource-based and dynamic-capability views of rent creation. Strateg Manag J 22(5):387–401
21. Swink M, Hegarty WH (1998) Core manufacturing capabilities and their links to product differentiation. Int J Oper Prod Manage 18(4):374–396

22. Vehtari S (2006) The dynamics involved with manufacturing capabilities towards a competitive advantage, PhD Thesis. University of Technology, Helsinki. 2006/1 ISBN 951-22-8390-5
23. Hayes RH, Upton DM (1998) Operations based strategy. California Manage Rev 40(4):8–24
24. Teece D, Pisano G, Shuen A (1997) Dynamic capabilities and strategic management. Strateg Manag J 18(7):509–533

Chapter 4
Evolving Paradigms in Manufacturing

Abstract As the focus and perspectives in manufacturing have developed over time, the changes have often been described as shifts in paradigms, "a set of beliefs that guides action". These paradigms are not absolute in terms of complete frameworks for how to conduct business or organize manufacturing. What they represent are more coherent sets of principles and methods that inspire companies. Strategies will often have elements from different paradigms even if it is claimed they have adopted only one. Defining paradigms is not an exact science; rather, to a large extent it is a question of choosing a set of criteria supporting the purpose of categorization. Craft manufacturing, mass manufacturing, lean manufacturing, mass customization and personalization and sustainable manufacturing are paradigms that are frequently referred to in the literature.

4.1 From Craft to Adaptive Manufacturing

As the focus and perspectives in manufacturing have developed over time, the changes have often been described as shifts in paradigms, "a set of beliefs that guide action" [1]. These paradigms are not absolute in terms of complete frameworks for how to conduct business or organize manufacturing. What they represent are more coherent sets of principles and methods that inspire and guide manufacturing systems design.

Strategies will often have elements from different paradigms even if it is claimed they have adopted only one. Paradigms and 'ruling principles' are often short lived in the literature. However, the time-lag between acceptance and implementation of new principles are not reasons to focus only on the latest paradigms. Paradigms guide action not only through their principles, but also through their 'mental room to maneuver', and are an important part of the strategic context of manufacturing. Further, paradigms represent basic approaches to

A. Rolstadås et al., *Manufacturing Outsourcing*,
DOI: 10.1007/978-1-4471-2954-7_4, © Springer-Verlag London 2012

Time	Level	Enabler			Driver	
Before 1970	Single workstation	Information		Design	Technology	Push
		Materials	Input			
		Energy				
		Transformation technology				
		Controls				Pull
		Means				
1970 to 1980	Group of workstations	Information		Imple-mentation	Society and market	Environment
		Materials	Input			
		Energy				
		Transformation technology				
1980 to 1990	Manufacturing logistics area	Controls				
		Means				
		Information		Use		Customization
		Materials	Input			
		Energy				
1990 to 2000	Facility general structure	Transformationtechnology				
		Controls				
		Means				
		Information		Recon-figuration Dismis-sion		Price
		Materials	Input			
		Energy				
After 2000	Production network	Transformation tech.				Innovation
		Controls				
		Means				

Fig. 4.1 Criteria for identifying paradigms [Jovane et al. 4]

different issues, but are also often a toolbox of methods and principles that can be used in the more practical aspects of strategic planning, such as those related to analysis and implementation. The term "operative paradigm" has been introduced by Arbnor and Bjerke [2] to indicate a toolbox of methods and tools that allows for the resolution of problems, given a certain methodological approach.

Defining paradigms is not an exact science; rather, to a large extent it is a question of choosing a set of criteria supporting the purpose of categorization. Paci et al. [3] have presented criteria in what they call a "Production Paradigms Ontology" (PPO) which focuses on knowledge- and innovation. PPO is based on the NEST (Nature, Economy, Society, Technology) context defined by Jovane et al. [4] and illustrated in Fig. 4.1.

Jovane et al. [4] use these criteria to identify and describe five manufacturing paradigms:

- *Craft Manufacturing*—which means to make exactly the product that the customer asks for, usually one product at a time in a "pull-type business model": sell (get paid)—design-make-assemble. The processes have a low level of automation, but use skilled and flexible workers
- *Mass Manufacturing*—means to manufacture high quantities of identical products, selling them to customers and markets that will absorb what is manufactured. High volume means low costs and cheaper products, thereby increasing the market, which implies a "push-type business model": design—make-assemble-sell. The moving assembly line is an enabler for this paradigm, requiring standardized processes and specialized workforces
- *Flexible manufacturing*—has been the answer to the increased complexity and uncertainty in the business environment. In the 1970s overproduction and the demand for more diversified products resulted in decreased lot size and the requirement for shorter time-to-market. The manufacturing still followed principles from mass manufacturing, but were more module-based in order to meet the demand for variation. The business model is a mixed "push–pull model": design–make–sell–assemble
- *Mass customization and personalization*—is a society driven paradigm due to customers asking for greater variety in products, and also as a result of globalization creating a huge excess of production capacity. This situation has put customers in power and the manufacturer aims to manufacture a variety of almost customized products at mass production prices. The business model is based on pull: sell—design-make-assemble
- *Sustainable manufacturing*—is based on societies' needs for 'clean' products and product-life-cycle management related to clean products. Nano-, bio- and material technology are regarded as enablers for sustainable manufacturing.

These paradigms are frequently referred to in the literature and they are all still relevant in manufacturing, even if they have been linked to certain periods in time. Sustainable manufacturing is a paradigm that has received much attention recently, for example in the IMS 2020 roadmap for a realistic and desirable future for manufacturing [5]. Geyer and Scapolo [6] describe sustainable manufacturing as interplay between social, environmental, economic, and technological factors. The point is to develop strategies that allow people to live better lives while consuming fewer resources. In this way, sustainable manufacturing could be combined with principles from other paradigms, such as lean manufacturing.

4.2 The Lean Paradigm

After the World War II, Japanese companies were developing an entirely new approach to manufacturing characterized by an emphasis on reliability, speed, Just-In-Time, and flexibility, rather than volume and cost. Taiichi Ohno [7] at Toyota brought all of these themes together into what became the Toyota

Production System. Subsequently, Womack et. al [8]. reviewed and extended the principles from Toyota Production Systems (TPS), and introduced the term "*lean*". Foster defines "lean" as "a productive system whose focus is on optimizing processes through the philosophy of continual improvement" [9, p. 87]. One crucial insight is that most costs are assigned when a product is designed. As a consequence, product development activities should be carried out concurrently, not sequentially, by cross-functional teams. At the system engineering level, requirements are reviewed with marketing and customer representatives to eliminate costly requirements.

It has been argued that the quality and cost management focus in lean manufacturing is in contrast to the focus on availability and market segments in flexible production [10]. However, through the contributions of Womack et al. [8], Liker [11], and others, the focus of lean manufacturing has been extended. Lean manufacturing represents more than principles related to the shop floor and streamlining of operations: it is also a philosophy about relations to stakeholders, empowering employees, etc., and as such is a reference for manufacturing in many industries. Lean manufacturing has basically the same characteristics as Jovane et al. [4] describe for flexible manufacturing, and will hereafter be referred to as the flexibility paradigm.

"*Agile manufacturing*," is a relatively new term that was first introduced with the publication of the Iacocca Institute report twenty-first century Manufacturing Enterprise Strategy in 1991. Agile manufacturing focuses on mass customization through processes, tools, and training that respond quickly to customer needs and market changes while still controlling costs and quality. The aim is to develop agile properties as a competitive advantage by being able to rapidly respond to changes occurring in the market environment and through the ability to use and exploit the knowledge-resource. There is focus on the deployment of advanced information technologies and the development of highly nimble organizational structures to support highly skilled, knowledgeable, and empowered people [12].

Adaptive manufacturing is closely related to agile manufacturing, but adds a higher level of automated adaptation of operations [13]. Klocke [14] emphasizes the following elements: the need for technological leadership, implementation of virtual process chain layouts, automated and high-productivity manufacturing systems, turnkey production system delivery, and network compatibility. Adaptive manufacturing has the characteristics of mass customization, and will be referred to as mass customization in the subsequent chapters.

4.3 Sustainable Manufacturing

Sustainability has gained much attention over the last decade within manufacturing. We now talk about a "sustainability manufacturing paradigm". This increased focus on sustainability is to a large extent explained by human activities, the natural environment and resources under stress.

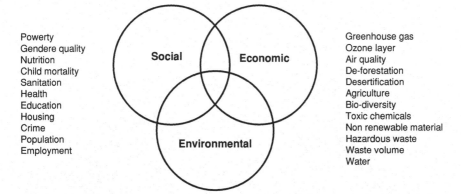

Powerty
Gendere quality
Nutrition
Child mortality
Sanitation
Health
Education
Housing
Crime
Population
Employment

Greenhouse gas
Ozone layer
Air quality
De-forestation
Desertification
Agriculture
Bio-diversity
Toxic chemicals
Non renewable material
Hazardous waste
Waste volume
Water

Fig. 4.2 United Nations indicators of "sustainability"

First defined by the Brundtland commission in 1987 [15], sustainable development is "development that meets the need of the present without compromising the ability of future generations to meet their own need". Sustainability and *Corporate Social Responsibility* (CSR) is frequently defined as the composition of three elements: environment, social and economic (see Fig. 4.2).

For the purposes of Commerce's Sustainable Manufacturing Initiative, sustainable manufacturing is defined by the US department of commerce as [16] "the creation of manufactured products that use processes that minimize negative environmental impacts, conserve energy and natural resources, are safe for employees, communities, and consumers and are economically sound".

Many manufacturing companies have been 'forced' to think social responsibility and sustainability through regulations, new standards or public pressure towards greener products. However, many companies have also approached sustainable manufacturing using a more visionary and proactive approach with the belief that their actions on these issues will make a difference. There are also many examples of manufacturing companies who has made CSR and sustainable manufacturing to the basis for their business models reflected in for example innovation policies, sales and marketing, as well as partner selection.

Industries have traditionally addressed pollution concerns at the point of discharge but this end-of-pipe approach is often costly and ineffective. Thus, industry has increasingly adopted cleaner production by reducing the amount of resources and energy used in the production process, and a focus on environmental friendly energy sources and materials. This also means that many companies now are considering the environmental impact throughout the product's lifecycle and are integrating environmental strategies and practices into their own management systems. While more integrated practices, such as closed-loop production, could potentially yield substantial environmental improvements, they can only be realized by combining a wide range of innovation targets and mechanisms, and both technological and non-technological changes. This is often referred to as *system innovation*.

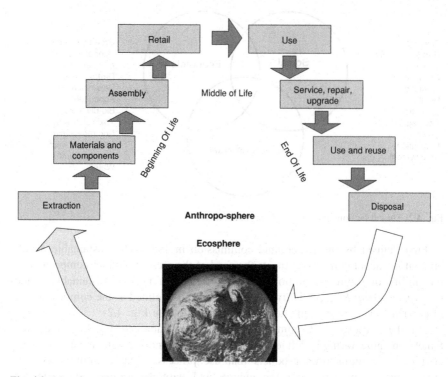

Fig. 4.3 Manufacturing in a life cycle perspective

Allwood [17] focuses on the relation between the "ecosphere" and "anthro-posphere" where sustainable manufacturing is about developing capabilities to transform materials without emission of greenhouse gases, use of non-renewable or toxic materials or the generation of waste. This means that sustainable manu-facturing should be seen in a life cycle perspective as illustrated in Fig. 4.3.

Life Cycle Analysis (LCA) has the goal to create knowledge and compare the full range of environmental effects assignable to products and services in order to improve processes, support policy and provide a sound basis for informed deci-sions. LCA methodologies aiming to assess environmental impacts associated with all the stages i.e. from raw material extraction through materials processing, manufacture, distribution, use, repair and maintenance, and on to disposal or recycling.

Sustainability could be combined with manufacturing principles from different paradigms. However, this requires that these principles are focused on resource efficiency and also where the environmental and social aspects of manufacturing become the premises for strategic, tactical and operational decisions.

References

1. Denzin N, Lincoln Y (1994) Introduction: entering the field of qualitative research. In: Denzin N, Lincoln Y (eds) Handbook of qualitative research. Sage Publications, London, pp 1–17
2. Arbnor I, Bjerke B (1997) Methodology for creating business knowledge. Sage Publications, Thousand Oaks
3. Paci MA, Chiacchio MS, Lalle C (2008) Productions paradigms ontology (PPO): a response to the need of managing knowledge in high-tech manufacturing. In: Bernard A, Tichkiewitch S (eds) Methods and tools for effective knowledge life-cycle-management. Springer, Heidelberg, pp. 227–240
4. Jovane F, Koren Y, Boer CR (2003) A present and future of flexible automation: towards new paradigms. Ann CIRP 53(1):543–560
5. Taisch M, Cassina J (2010) Action roadmap on key area 1, 2 and 3. IMS2020 Project Report, POLIMI, Italy
6. Geyer A, Scapolo F (2004) European manufacturing in transition—the challenge of sustainable development: Four scenarios 2015–2020. Innov Manage Policy Pract 6:331–343
7. Ohno T (1988) Toyota production system: beyond large-scale production. Productivity Press Inc, Cambridge
8. Womack JP, Jones DT, Roos D (1990) The machine that changed the world: the story of lean Production. Harper Business, New York
9. Foster TS (2006) Managing quality–integrating supply chain. Prentice Hall, New Jersey
10. SAP AG (2003) Manufacturing strategy: an adaptive perspective. SAP White Paper mySAP SCM. Available at http://www.sap.com/industries/automotive/pdf/BWPWP_Manufacturing_Strategy.pdf. Accessed April 2007
11. Liker J (2004) The toyota way: 14 managament principles from the world's greatest manufacturer. McGraw-Hill, New York
12. Goldman L, Nagel R, Preiss K (1995) Agile competitors and virtual organizations–Strategies for enriching the customer. Van Nostrand Reinhold, New York
13. Røstad CC, Stokland Ø (2007) Adaptive manufacturing from enterprise to shop-floor level—sense, determine and execute. SINTEF Technology and Society, SINTEF rapport, STF50 A07035, Trondheim
14. Klocke F (2004) Adaptive manufacturing, The MANUFUTURE 2004 workshop, Fraeuenhofer IPT, Dortmund, July 2004. http://e.europa.eu/research/industrial_technologies/articles/article_1292_en.html. Accessed Dec 2008
15. Brundtland GH (1987) Our common future. World Commission on Environment, Development. Oxford University Press
16. US department of commerce (http://trade.gov). Accessed Mar 2011
17. Allwood J (2005) What is sustainable manufacturing? Sustainable manufacturing seminar series, 16th February 2005. Institute for Manufacturing. University of Cambridge

Chapter 5
The Knowledge Dimension

Abstract The capabilities of organisations are closely related to knowledge. Resources can be regarded as capabilities if we have the knowledge to exploit them. The development of technological, but also human and organizational capabilities is important to create, transfer and adapt knowledge. These capabilities are often defined as a companies' 'infrastructure' as opposed to 'structure' which include the more tangible aspects such as physical assets (facilities and capacities), vertical integration and process technology. Typically, it is the transfer and combination of knowledge that initiate changes for example innovations and improvements. Knowledge creation is a difficult process, involving people and their knowledge in a way that not only supports knowledge creation, but also accelerates it. Knowledge transfer across organizational boundaries can involve tacit, explicit, and cultural knowledge to varying degrees. Being global means companies have opportunities for developing a wider repertoire of knowledge, but it will be more difficult and costly to transfer the knowledge to other corporate units that do not possess the same relational absorptive capacity.

5.1 Capabilities Developed Through Knowledge

To gain a good position in the market place companies must be capable of fulfilling customer requirements better or at least as well as their competitors. In other words the companies' capabilities are essential for success and from the definitions of manufacturing strategy we saw earlier, it is about developing capabilities. But what does the term capability mean?

Capabilities are closely related to knowledge. Resources can be regarded as capabilities if we have the knowledge to exploit them. Technological, but also human and organizational aspects are important to create, transfer and adapt knowledge to develop capabilities. These aspects are often defined as a companies'

A. Rolstadås et al., *Manufacturing Outsourcing*,
DOI: 10.1007/978-1-4471-2954-7_5, © Springer-Verlag London 2012

'infrastructure' in contrast to 'structure' which are the more tangible aspects of manufacturing such as physical assets (facilities and capacities), vertical integration and process technology.

Skinner [1] asserts the importance of human factors right from the beginning of development i.e. how a company develops the policies and systems that govern such activities as capital budgeting, human resources, quality/process control, material flows, and performance measurement are more important than structural decisions. He argues that such systems should be designed to encourage the continual adaptation and improvement of an organization's skill base rather than to achieve some 'optimal' strategic fit.

Infrastructure or 'soft' decisions relate to people, organizations and systems, and support the action element of a strategy, typically involving the time of middle managers on a more frequent basis than decisions around structure. Walker [2], Bartlett and Ghosahl [3] and others emphasize the coordination challenge between, product, functional, and geographically oriented management when operating globally.

5.2 Knowledge, the Basis for Innovations and Improvements

In the literature there has been little dispute over the importance of knowledge creation in developing capabilities and exploiting opportunities. There are fewer contributions on how to initiate and coordinate the knowledge creation processes. A key issue has been that knowledge is almost always dispersed among many people. Hayek [4] observed in 1945 that knowledge is not, and cannot, be concentrated in a single mind, and no single mind can specify in advance what is relevant knowledge in a particular context. Typically it is the transfer and combination of knowledge that initiate changes, innovations and improvements.

Knowledge creation is a difficult process, involving people and their knowledge in a way that not only supports knowledge creation, but also accelerates it. Argyris [5] has discussed the problem, focusing on extending cycles of learning from individual level to organizational level. Knowledge creation is to a significant degree context specific or even relation specific [6, 7]. *For instance, extensive long-term cooperation with a specific customer or supplier will enhance the subsidiary's absorptive capacity, its problem-solving capacity, and its ability to create new knowledge within that context* [8, p. 444]. However, it will be more difficult and costly to transfer the knowledge to other corporate units that do not possess the same relational absorptive capacity.

Knowledge transfer is an imperative for knowledge issues, whether they are related to R&D or day-to-day operations. Knowledge creation is about increasing an organization's knowledge base (analytic and synthetic). This means that the transfer of knowledge from outside and also between units and people is a basic prerequisite. In 1994 Nonaka introduced a theory for the coordination of organizational knowledge creation processes that has received much attention:

Fig. 5.1 The knowledge
creation spiral [10]

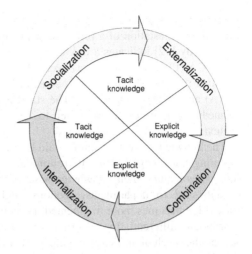

"Its central theme is that organizational knowledge is created through a continuous dialogue between tacit and explicit knowledge" [9, p. 14]. The nature of this dialogue includes four patterns of interaction of tacit and explicit knowledge: socialization, externalization, combination, and internalization.

Externalization is defined as "the process of articulating tacit knowledge into explicit concepts via such means as metaphor, analogy, hypotheses or models" [10, p. 67]. Internalization is defined as "the process of embodying explicit knowledge into tacit knowledge through socialization" [10, p. 69]. Socialization involves the sharing of knowledge between individuals, while combination is the conversion of explicit knowledge into more complex sets of explicit knowledge, thereby fitting the elements of knowledge together.

Knowledge transfer across organizational boundaries can involve tacit, explicit, and cultural knowledge to varying degrees. Nonaka and Takeuchi [10] identify a knowledge transfer process whereby an individuals' knowledge creation that takes place in groups may be shared with the whole organization. Figure 5.1 illustrates the different modes of knowledge transfer, and the knowledge creating process going through the different modes, wherein each mode creates different forms of knowledge and is initiated by different triggers.

Knowledge transfer requires a substantial amount of collateral knowledge on the part of the receiving organization to be able to decode and apply it. Nonaka and Konno [11] emphasize the role of "ba", which represents a context or a shared space for emerging relationships [11]. There are different types of "ba":

- *Physical ba*—offices, dispersed business spaces, etc.
- *Virtual ba*—e-mails, teleconferences, etc.
- *Mental ba*—shared experiences, ideas, ideals, etc.

These types have differing importance and roles in the knowledge creation process. During product development the physical ba may be important in an initial phase where product ideas are shared and possibly physical prototypes are developed. Virtual ba may be more important when there is a common picture, a physical manifestation of a products or process.

Kennedy [12] indicates a link between the knowledge creation spiral [10] and the PDCA (Plan-Do-Check-Act) cycle [13]. He sees PDCA as a learning process, in which the plan (P) and do (D) constitute a learning phase, and check (C) and act (A) are about standardizing. The standardizing of knowledge is, according to Kennedy [12], necessary in order to be able to use it in another project or context.

The knowledge creation spiral could be criticized for not being dynamic, for not capturing knowledge outside its routine activities. Bathelt et al. [14] see that as companies mature, their knowledge base will grow unevenly. When the expertise is addressed to departments, sections, and geographical units, the body of knowledge becomes more fragmented. Bathelt et al. [14] argue that by creating an appropriate differentiation, a group of companies can develop knowledge far beyond the reach of any single member. This view is also supported by Simon who describes how divergent knowledge bases may give rise to new types of input and new way of combining knowledge [15].

Lee and Cole [16] have proposed a knowledge creation model which challenges the traditional company-oriented approach. Their community-based model has adapted the evolutionary framework, emphasizing the role of critical evaluation as a key driver [17]. Lee and Cole [16] use the Linux kernel operating system as a reference for their analysis.

Manufacturing paradigms have inherent principles and guidelines for how 'things are done'–problem solving, stakeholder relations, and so on. This is typically reflected in innovation processes. Lundvall [18] distinguishes between interactive and linear models of innovation. A linear model has well-defined sequences and tasks within, for example, R&D projects. The linear model will often rely on analytical knowledge, while the 'interactive model' is more relevant in incremental innovations. An interactive learning process between practitioners and experts (e.g. an R&D department) will normally rely on both a synthetic and an analytical knowledge base. The innovation process is much more decentralized, interactive and incremental within lean manufacturing than in mass manufacturing. More people, and different people, are involved in the lean knowledge creation process than in the more linear, formalized, and centralized innovation processes we see in mass manufacturing.

The synthetic knowledge base is very important for innovations in lean manufacturing, whereas the centralized innovation process in mass manufacturing is often more radical, depending on the analytical knowledge base. Craft manufacturing is extremely decentralized and dependent on the synthetic knowledge base, to some extent explained by its size, but the lack of standardization makes problem solving and incremental innovation a part of the 'daily routine'. Within adaptive manufacturing the innovation processes are often very complex, with companies dependent on profiting from common knowledge properties and innovations within networks. This requires continual improvements and incremental innovations, but also participating in radical innovations, if required within the network.

Being global means companies have opportunities for developing a wider repertoire of knowledge. However, the global context represents a spatial and socio-cultural dimension which makes coordination and knowledge transfer even more challenging.

Since knowledge is the basis for developing capabilities that are prerequisites in the market place, strategic manufacturing decisions, in particular related to outsourcing and partner selection, need to highlight the impact on knowledge transfer of these decisions. Manufacturing companies that are able to be part of knowledge creation processes related to the drivers for change in manufacturing will also be in a position to develop capabilities that makes them winners in the market place.

References

1. Skinner W (1969) Manufacturing—missing link in corporate strategy. Harward Bus Rev 47(3):136–145
2. Walker J (1996) Redemption revival. Marketing Week 19(7):67–73
3. Bartlett CA, Ghosal S (1992) Transnational management, text, cases, and readings in cross-border management. Irwin, Chicago
4. Hayek HF (1945) The use of knowledge in society. Am Econ Rev 35(4):519–530 Sept
5. Argyris C (1990) Overcoming organizational defenses. Facilitating organizational learning. Allyn and Bacon, Boston
6. Lane PJ, Lubatkin M (1998) Relative absorptive capacity and interorganizational learning. Strateg Manag J 19(5):461–477
7. Forsgren M, Johansson J, Sharma D (2000) Development of MNC centres of excellence. In: Holm U, Pedersen T (eds) The emergence and impact of MNC centres of excellence. Macmillan, London, pp 45–67
8. Bjorkman I, Barner-Rasmussen W, Li L (2004) Managing knowledge transfer in MNCS: the impact of headquarters control mechanisms. J Int Bus Stud 35:443–455
9. Nonaka I (1994) A dynamic theory of organizational knowledge creation. Organizational Sci 5(1):14–37
10. Nonaka I, Takeuchi H (1995) The knowledge creating company. Oxford University Press, New York
11. Nonaka I, Konno N (1998) The concept of "ba": building a foundation for knowledge creation. California Manage Rev 40(3):40–54
12. Kennedy MN (2010) Knowledge based product development—understanding the true meaning of lean in product development. Presentation at the seminar Knowledge based development forum, Kongsberg, 27–28 Jan 2010
13. Deming WE (1986) Out of crisis Boston. MIT/CAES, Cambridge
14. Bathelt H, Malmberg A, Maskell P (2004) Clusters and knowledge: local buzz, global pipelines and the process of knowledge creation. Prog Hum Geogr 28(1):31–56
15. Simon HA (1985) What we know about the creative process. In: Kuhn RL (ed) Frontiers in creative and innovative management. Ballinger, Cambridge, pp 3–20
16. Lee GK, Cole RE (2003) From a firm-based to a community-based model of knowledge creation: the case of the linux kernel development. Organizational Sci 14(6):633–649
17. Popper K (1972) Objective Knowledge, revised edition 1989. Oxford University Press, New York
18. Lundvall BÅ (1992) National systems of innovation: towards a theory of innovation and interactive learning. Pinter Publishers, London

Part II
The Engine Driving Industrial Change

There is a continuous increase in demand for goods in contrast with a diminishing availability of resources. Resource availability and environmental impact represent constraints in the development of products and manufacturing processes. Simultaneously, we have been heading into a global economic situation that is less predictable and stable than before. This gives rise to challenges in how to organize the manufacturing system in the future and how to improve the transformation process. This part of the book describes how the global competitive landscape is changing, and where strategies towards innovation and knowledge are becoming increasingly important.

Part II
The Engine Driving Industrial Change

Chapter 6
Industrial Outlook

Abstract The lack of effective and inclusive global governance on issues such as, financial stability, trade, climate change and security will be a source of increasing risks in the next decade. Manufacturing continuously meets regional economic and/ or political crises. In addition, more global economic downturns, as experienced throughout 2008 and 2009, have challenged the established industry. These crises often tend to cause a society to restructure and 'renew' industries where some companies fail because they don't have the right strategies or abilities to survive while other companies succeed by finding new markets and business opportunities based on new competencies. Coming from on a variety data sources, 'The Global Competitiveness Report' published by the World Economic Forum is a comprehensive assessment of more than 130 economies. Fortunately, the picture of potential 'winners' and 'losers' in the world of manufacturing is changeable. There will normally be political, cultural and other external factors in the business environment, that when combined with factors and decisions taken by companies, organizations and individuals can alter the competitive landscape.

6.1 Manufacturing in Turbulent Periods

Manufacturing continuously meets regional economic and political crises. The global economic downturns experienced throughout 2008 and 2009 have challenged the established industry. These crises often tend to restructure and 'renew' industries where some companies who don't have the right strategies or abilities to survive stagnate and decline, while others grow, find new markets and business opportunities.

Companies that think they just can reduce their activities or turn down factory output for a while and then later pick up where they left off without renewal, normally end up in an even worse situation. The ones that normally come out of crisis as the winners are those that use the downturns to rethink what they are doing; meeting new customer needs, adjusting their manufacturing processes and rethinking their product offerings. After a crisis like the recent global financial crisis, not only do we see changes to customer behaviour, but we also see new technologies, products and business processes, and new more agile competitors in the market place.

The normalization and stability to which the global economy seemed to have headed after reaching the trough of a steep downturn in 2009, has been broken once again by a troubling string of events and data reports. Such reports tend to contradict each other and it is not easy to get a good picture of the new reality for industry and current trends in manufacturing. Companies can't rely only on input and conclusions drawn by others, they need to have their own ways of dealing with such knowledge so that the specific context of the company is the premise for making strategic decisions.

The business environment is changing more rapidly than ever before where old governance structures created for demands of an earlier and different time are struggling with the new challenges. The lack of effective and inclusive global governance on issues such as: financial stability, trade, climate change and security will be a source of increased risks towards the next decade. However, we see that trading blocs such as the EU, and developing economies such as BRIC countries (Brazil, Russia, India, China) and phenomena such as Wikileaks and Facebook that did not exist in the governance structures from a post-World War setting, are now beginning to shape agendas worldwide [1].

6.2 Technology Outlook

Trends in the business environment described in the previous chapters are premises for manufacturing and set directions for company specific strategies as well as for nations and trading blocks. Some of these trends represent challenges that motivate and trigger innovations and improvements towards sustainability and for innovations in technology and the way manufacturing is organized. Technology itself represents and important engine for change and innovation. Cheaper, smaller, more powerful computers, plus increased wireless connectivity are becoming key features in many products and in many manufacturing processes. 'Moore's law' will still be valid in the next decade. Moore's law refers to how Intel co-founder Gordon Moore in 1965 noticed that the number of transistors per square inch on integrated circuits had doubled every year since their invention. He predicted that this trend would continue into the foreseeable future [2].

Information and communication technology has had an enormous positive impact on personal life, business, and society at large, through easy production and sharing of information. But the exponential growth in data not only results in greater automation and ubiquitous computing, but also raises retrieval and security issues.

A major aspect of ICT as a driver for innovations is connectivity. Connectivity that encompasses Internet access, mobile telephony, and all kinds of gadgets with wireless connections that also leaves behind a trail of digital traces. These traces could be important inputs for knowledge about customers and product requirements. An ever increasing number of products containing embedded software are also a result of the higher product flexibility, adaptivity, and robustness, to reduce cost and time-to-market. Digital control systems are replacing mechanical control in a world where scalable product version management, release control, verification and validation are rising rapidly [1].

Companies, nations and regions that manage to take the most advantage of these technologies will be well positioned to be leading industrial players in the years to come. The IMS 2020[1][3] initiatives highlights examples of how nations, regions and companies in collaboration can define common visions for a realistic and desirable future for manufacturing. This vision focuses on sustainability and technology as an enabler for efficient, customer oriented manufacturing. A roadmap has been developed for the development of research and education to create knowledge about future manufacturing.

6.3 Industrial Production: Who Takes the Lead?

A MAPI (Manufacturers Alliance) report from October, 2010 [4] envisions that in the near-term global growth prospects will be dominated by a halting and uncertain rebound in advanced economies. This, in turn, is already muting the strength of demand for exports, a critical and often leading component of developing country growth. Gross domestic product (GDP) in non-U.S. industrialized countries, which include Canada, the Eurozone and Japan, is expected to slow from a compound annual rate of 2.6 percent during the second quarter of 2010 before reaching a trough of 1.9 percent during the first half of 2011. Following that, MAPI sees industrialized country growth advancing to 2.2 percent during the second half of 2011. The volatility that has plagued currency markets in recent years are also expected to continue.

An important question is whether this snapshot of the economic situation in 2010 showing volatility and uncertain growth perspectives are just the final phase of the downturn, and that the situation will normalize soon. An alternative scenario could be that we now are heading into a more complex and shifting business competitive landscape. This would require an even stronger emphasize on knowledge and flexibility.

We have also seen that not all parts of the globe were hit equally hard by the global financial crisis. One example is from South America where countries such

---

[1] IMS 2020 is a research project sponsored by the European Union defining a roadmap for intelligent manufacturing research.

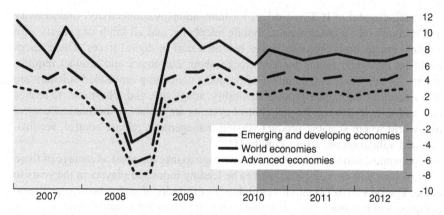

Fig. 6.1 Global GDP growth (IMF 2010) [6]

as Argentina have to some extent strengthened their position during this period. While Argentina's GDP grew 0.9% in 2009 according to Argentine government data [5], their industrial production expanded by as much as 11.8% in the first seven months of 2010. This has to a large extent been explained by the strong growth by top trade partners Brazil and China. Heavy government spending and the weak currency are helping to fuel a strong expansion in the economy. IMF (International Monetary Fund) estimates also support the emergence of these developing countries with improved competitiveness (Fig. 6.1).

Developing economies have for the most part fared comparatively well during the crisis. Brazil, China, and India are expected to grow at rates of between 5.5 and 10 percent in 2010, with growth holding up well over the next few years. According to Klaus Schwab, the Executive Chairman of the World Economic Forum, the world increasingly looks to the developing world as the major engine of growth in the global economy [7]. According to the MAPI report, the developing countries will likely outperform their more established industrialized counterparts in regions such as United States and Western Europe.

Fortunately the picture of winners and losers in the world of manufacturing is not necessarily unchangeable. This means that companies, nations and regions can change the scene if they are proactive, use the right combination of incentives and make robust strategies to meet the manufacturing demands of the future.

6.4 What Defines the Winners?

It is not evident to identify who will become leading countries and regions in manufacturing in the future. The concept of competitiveness involves static and dynamic components. There will normally be a synthesis of political, cultural and other external factors in the business environment, in combination with decision

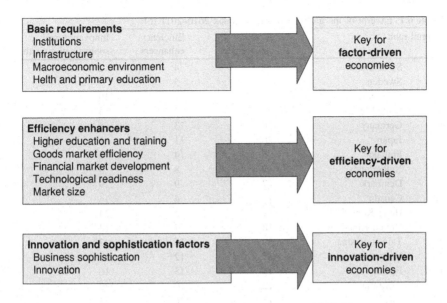

Basic requirements
Institutions
Infrastructure
Macroeconomic environment
Helth and primary education

Key for
factor-driven
economies

Efficiency enhancers
Higher education and training
Goods market efficiency
Financial market development
Technological readiness
Market size

Key for
efficiency-driven
economies

Innovation and sophistication factors
Business sophistication
Innovation

Key for
innovation-driven
economies

Fig. 6.2 Pillars of competitiveness (global competitiveness report 2010) [6]

making processes taken by companies, organizations and individuals. This complexity is illustrated in Fig. 6.2.

Based on a variety data sources The Global Competitiveness Report (Table 6.1) [7] published by the World Economic Forum is a comprehensive assessment of more than 130 economies (2010) according to over 100 indicators grouped under the competitiveness pillars in Fig. 6.2. The report shows that overall Europe continues to feature prominently among the most competitive regions in the world. This report gives important input to any companies strategy making process around issues such as evolving markets and in particular locations for manufacturing units and strategic alliances.

Switzerland retains its 1st place position, characterized by a high capacity for innovation and a sophisticated business culture. Public institutions in Switzerland are among the most effective and transparent in the world. They have an excellent infrastructure, a well-functioning goods market, and a highly developed financial market as well as a labor market that is among the most efficient in the world. Sweden (2nd) benefits from the world's most transparent and efficient public institutions, with very low levels of corruption. Private institutions also receive excellent marks with companies demonstrating ethical behavior, strong auditing and reporting standards, and well-functioning corporate boards. Goods and financial markets are also very efficient. A strong focus on education and a high level of technological adoption and a very sophisticated business culture have made Sweden one of the world's leading innovators.

The German (5th) macroeconomic environment has improved compared with other advanced economies and gets a high score for quality of transport and

Table 6.1 Excerpt of the global competitive index 2010–2011 [6]

Overall rank		Basic requirements	Efficiency enhancers	Innovation and sophistication factors
1	Switzerland	2	4	2
2	Sweden	4	5	3
3	Singapore	3	1	10
4	United States	32	3	4
5	Germany	6	13	5
6	Japan	26	11	1
7	Finland	5	14	6
8	Netherlands	9	8	8
9	Denmark	7	9	9
10	Canada	11	6	14
11	Hong Kong SAR	1	2	24
12	United Kingdom	18	7	12
13	Taiwan, China	19	16	7
14	Norway	17	12	17
15	France	16	15	16
16	Australia	12	10	22
17	Qatar	13	26	23
18	Austria	15	19	13
19	Belgium	22	17	15
20	Luxembourg	10	20	19
21	Saudi Arabia	28	27	26
22	Korea, Rep.	23	22	18
23	New Zealand	14	18	28
24	Israel	39	23	11
25	United Arab Emirates	8	21	27
26	Malaysia	33	24	25
27	China	30	29	31
28	Brunei Darussalam	20	67	72
29	Ireland	35	25	21
30	Chile	37	35	44
31	Iceland	41	31	20
32	Tunisia	31	50	34
33	Estonia	25	34	45
34	Oman	24	48	47
35	Kuwait	36	68	60
36	Chech Republic	44	28	30
37	Bahrain	21	33	55
38	Thailand	48	39	49
39	Poland	56	30	50
40	Cyprus	29	36	36
41	Puerto Rico	43	40	29
42	Spain	38	32	41

(continued)

Table 6.1 (continued)

Overall rank		Basic requirements	Efficiency enhancers	Innovation and sophistication factors
43	Barbados	27	52	52
44	Indonesia	60	51	37
45	Slovenia	34	46	35
46	Portugal	42	43	39
47	Lithuania	52	49	48
48	Italy	46	45	32
49	Montenegro	45	64	56
50	Malta	40	47	46

communication, efficient goods market, and sophisticated businesses. German businesses are also aggressive in adopting technologies for productivity enhancements while on the other hand the labor market is rigid and a hindrance to job creation. United Kingdom (12th) benefits from the efficiency of its labor market, sophisticated and innovative businesses that adapt to the latest technologies for productivity improvements, and operating in a very large market. The macroeconomic environment remains the country's greatest competitive weakness.

According to the The Global Competitiveness Report 2010 [7] a lack of macroeconomic stability is the United States' greatest area of weakness. However, US companies are highly sophisticated and innovative, supported by an excellent university system that collaborates strongly with the business sector in R&D. Combined with the scale opportunities afforded by the sheer size of its domestic economy these qualities continue to make the United States very competitive (4th).

The Russian Federation (63rd) is in a competitive position where the macroeconomic instability has been somewhat balanced by improvements in other areas, such as education, as well as technological readiness. Inefficiencies in goods markets reduce the country's ability to take advantage of some of its strengths such as a high innovation potential and scores well on higher education and training. A particular challenge for Russia is related to insufficient protection of property rights, undue influence, and weak corporate governance standards.

The competitiveness assessment for Latin America and the Caribbean shows that important progress has been made by several countries in improving and reinforcing their competitiveness fundamentals. While Bolivia, Panama, and Paraguay post the largest improvements, many other regional economies improve slightly. The overall positive development is mainly explained by sounder fiscal management, increased market efficiency and openness, and export diversification, among other areas. However, a major challenge in the region is the quality of educational systems at all levels.

Singapore maintains its position at 3rd place, still the highest-ranked country from Asia. The country's institutions continue to be assessed as the best in the world. The country also get high scores for the efficiency of its goods-, labor- and financial markets. Singapore also has world-class infrastructure and has a strong focus on education. Japan (6th) has a major competitive edge in the areas of business sophistication and innovation and company spending on R&D. The country's overall competitive performance, however, is characterized by its macroeconomic weaknesses, with high budget deficits over several years. China (27th) has its main strengths in its large and growing market size, macroeconomic stability, and relatively sophisticated and innovative businesses, and has also an improved assessment of its financial market. Technological readiness is an area where China has traditionally underperformed but has improved while there remains considerable room for improvement in the quality of higher education and training.

Africa has experienced impressive growth over the past decade and has weathered the recent global economic downturn relatively well. Some African countries continue to fare quite well and South Africa (54th) and Mauritius (55th) remain in the top half of the rankings, and there have been measurable improvements across specific areas in a number of other African countries. However, generally, sub-Saharan Africa as a whole lags behind the rest of the world in competitiveness.

These rankings should be approached with caution, but they could for example indicate regional and country attractiveness according to certain location criteria for global companies. We could also get indications of trends and which countries and regions are succeeding in improving competitiveness. Trends and scores from the World Economic Forum, Global Competitiveness Index are supported by The Deloitte "2010 Global Manufacturing Competitiveness Index" [8]. This index is based on the responses of more than 400 chief executive officers and senior manufacturing executives worldwide to a survey conducted in late 2009 and early 2010. This report also gives anticipated projections for future competitiveness. The Deloitte report identified the emergence of a new group of leaders in the manufacturing competitive index over the next years. These include Mexico, Poland, Brazil and Russia. Not unexpectedly, Asian giants like China, India, and the Republic of Korea are projected to continue to dominate the index in the years to come. Furthermore, dominant manufacturing super powers of the late 20th century; the United States, Japan, and Germany are expected to become less competitive over the next five years.

The complexity of the interplay between factors influencing competitiveness is illustrated by how they are weighted differently in the various global regions. For example in Europe 'talent-driven innovation' is ranked as the most important driver followed by "energy cost and policies" and "quality of physical infra-structure", while in Asia "talent-driven innovation" is followed by "government's investments in manufacturing and innovation" and "cost of labor and materials".

The Deloitte report [8] also indicates that access to talented workers capable of supporting innovation is the key factor driving global competitiveness at manufacturing companies—well ahead of classic factors typically associated with competitive manufacturing, such as labor, materials, and energy.

References

1. Det Norske Veritas (2010) Technology outlook 2020. DNV Research & Innovation, Høvik
2. Moore GE (1965) Cramming more components onto integrated circuits. Electronics 39:1–4
3. IMS 2020 (2010) Roadmap on innovation, competence development and education. Accessed 15 July 2010
4. The Manufacturers Alliance/MAPI Global Report—Oct 2010 (ER-708) http://en.mercopress.com. Accessed Sept 2010
5. The Economist (2010) Europeans_want_dynamic_economy. http://www.economist.com/blogs/charlemagne/2010/01/do. Accessed Mar 2011
6. IMF (2010) World economic update. International Monetary Fund
7. Global Competiveness Report 2010–2011 (2010) World economic forum
8. A report by Deloitte's global manufacturing industry group and the U.S. council on competitiveness. http://www.deloitte.com/view/en_GX/global/industries/manufacturing/. Accessed Jan 2010

Chapter 7
Indicators and Initiatives for Industrial Renewal

Abstract There are a complex set of factors for improving the competitive position for industrialized countries and chief among them is an improved educational system. Improved education provides us with the new 'winners' in the global market place. In the late 1990s, developing countries began to recover some of the educational ground lost in the 1980s, when enrolments stagnated. Educational systems must fit the increased need for change and continuous learning in industry. Governments and policymakers need to continuously increase industry competitiveness. In Europe, for example important initiatives where launched under the heading of the "The Lisbon agenda" that included a range of initiatives for improvement across all sectors of industry. The term "learning economy" is frequently used in characterizing the current phase of socio-economic development in industrialized economies. The term illustrates the dynamic aspects of the economy, how companies needs to innovate and continuously adapt to and benefit from a changing business environment.

7.1 Education for Growth and Industrial Renewal

There are a complex set of factors explaining the improved competitive position for countries. Indicators for growth and attractiveness include natural resources, market access, governmental incentives, and so on. Improved educational levels in the population can make a significant difference to competitiveness. In the late 1990s, developing countries began to recover some of the educational ground lost in the 1980s, when enrolments stagnated. This is illustrated in Fig. 7.1.

The pace of progress has accelerated since 2000 and if trends between 2000 and 2008 continue, the increase in school-life expectancy in the current decade will be three times the level achieved in the 1970s. UNESCO (United Nations Educational, Scientific and Cultural Organization) statistics illustrates how primary education has reached much more of the world's children [1]. UNESCO

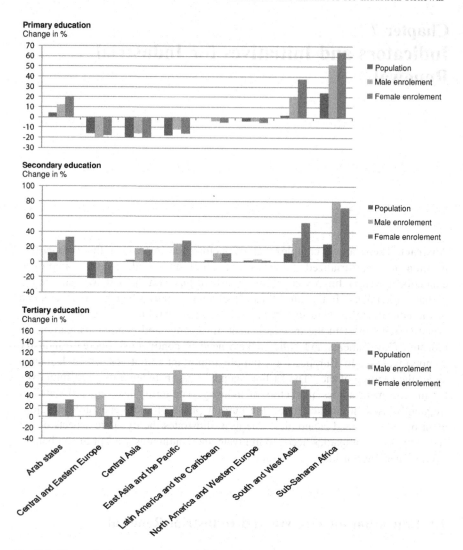

Fig. 7.1 Change in enrolment and population, 1999–2008 [1]

statistics in Fig. 7.2 show that the global average of expected years of schooling grew from 7.9 years in 1970 to 11.0 years in 2008. In sub-Saharan Africa, the value nearly doubled from 4.4 to 8.4 years. Despite this progress, the region has the lowest number of school years—almost half of the number of years in North America and Western Europe. The gaps are being reduced, but slowly.

The average educational level in a country and region is important. Also important are the educational structure and systems supporting teaching. The quality of educational systems and extent to which they are adapted to the future needs of an economy is an important feature of industrial growth.

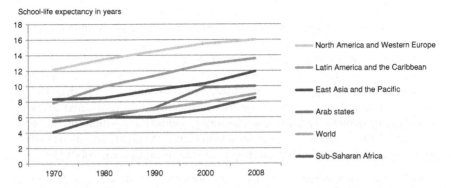

Fig. 7.2 School-life expectancy in years by region, 1970–2008 (UNESCO, 2010)

Educational systems must fit the increased need for change and continuous learning. According to Andreas Schleicher, OECD (Organization for Economic Co-operation and development) Education Directorate, we live in a fast-changing world, and producing more of the same knowledge and skills will not suffice to address the challenges of the future [2].

> A generation ago, teachers could expect that what they taught would last their students a lifetime. Today, because of rapid economic and social change, schools have to prepare students for jobs that have not yet been created, technologies that have not yet been invented and problems that we don't yet know will arise.

Education today is much more about ways of thinking which involve creative and critical thinking approaches to problem-solving and decision-making. It is also about ways of working, including communication and collaboration, as well as the tools they require, such as the capacity to recognize and exploit the potential of new technologies, or indeed, to avert their risks. Last but not least, education is about the capacity to live in a multi-faceted world as an active and engaged citizen. These citizens influence what they want to learn and how they want to learn it, and it is this that shapes the role of educators.

7.2 The Lisbon Agenda

Governments and policymakers around the globe attempt to continuously increase their industries competitiveness. In Europe important initiatives where launched during the millennium shift and addressed under the heading 'The Lisbon agenda' Although the prospects for the European economies were positive in 2000, European politicians and economists were looking for a response to China's official economic growth rate of 7.8% in 1998 and 7.1% in 1999. The rhetoric focused on the need for an answer to Chinas growth as well as a plan that could sustain Europes levels of the welfare. In the philosophy of competing states, in which a comparative advantage is used to generate economic growth, this meant creating a new strategy based on Europe's current primary resource-human capital.

When the heads of the European States met in 2000 they had high expectations of how they could develop a policy where innovation was seen as the motive for economic change. The strategy therefore set an array of policies and reforms designed to make Europe more competitive and innovation friendly. Knowledge and developing a dynamic knowledge based economy were considered a key to strengthen innovation and the manufacturing industries position in the new millennium. The European Council decided to launch the highly ambitious Lisbon Strategy. The minutes from the meeting read (European Council 2000) [3]:

> The Union has today set itself a new strategic goal for the next decade: to become the most competitive and dynamic knowledge-based economy in the world, capable of sustainable economic growth with more and better jobs and greater social cohesion

In order to achieve economic growth, a very broad reform package was proposed, ranging from implementing regulation on the Union's internal market to investing in a better infrastructure. In 2004, the existing achievements of the Lisbon Strategy were assessed stating that economic activities had not yet grown and more action was needed from all the actors in the European Union, both from the Member States and the European institutes. The Strategy was re-launched in March 2005 focusing on 5 key areas:

1. *Knowledge society*—increasing Europe's attractiveness for researchers and scientists, making R&D a top priority and promoting the use of information and communication technologies (ICTs)
2. *Internal market*—completing the internal market for the free movement of goods and capital, and taking urgent action to create a single market for services
3. *Business climate*—reducing the total administrative burden; improving the quality of legislation; facilitating the rapid start-up of new enterprises; and creating a more supportive environment for businesses
4. *Labour market*—delivering rapidly on the recommendations of the European Employment Taskforce; developing strategies for lifelong learning and active ageing; and underpinning partnerships for growth and employment
5. *Environmental sustainability*—spreading eco-innovations and building leadership in eco-industry; pursuing policies which lead to long-term and sustained improvements in productivity through eco-efficiency.

Opinions are divided in how successful 'Lisbon' has been and what should come next. The global competitive index in Fig. 6.1 indicates that the policy might not have had the desired effect. However, in retrospective it is difficult to say what had been the situation if the Lisbon agenda had not been developed.

7.3 Innovation Driving Industrial Change

We find innovations in most areas, in products, processes, services, organizations, social life, and so on. Innovations in one field also tend to influence other fields and trigger further innovation. One example is interactive social media such as Facebook which based on new software and business models has innovated social life, but has also increased markets for computers, touch screen devices, smart phones and similar devices. We often also see that new innovative products require innovations in manufacturing processes to reach customers with products at an acceptable price. OECD defines four types of innovations identified in the Oslo Manual for measuring innovation [4]:

- *Product innovation*—good or service that is new or significantly improved. This includes significant improvements in technical specifications, components and materials, incorporated software, user friendliness or other functional characteristics. In the education sector, a product innovation can be a new or significantly improved curriculum, a new educational software, etc
- *Process innovation*—new or significantly improved production or delivery method. This includes significant changes in techniques, equipment and/or software. In education, this can for example be a new or significantly improved pedagogy
- *Marketing innovation*—new marketing method involving significant changes in product design or packaging, product placement, product promotion or pricing. In education, this can for example be a new way of pricing the education service or a new admission strategy
- *Organizational innovation*—introducing a new organizational method in the firm's business practices, workplace organization or external relations. In education, this can for example be a new way organization of work between teachers, or organizational changes in the administrative area. These innovations can be new to the firm/educational institution, new to the market/sector or new to the world.

Even a brief look into the topic of innovation reveals a variety of different, mostly complimentary, definitions of innovation. A theoretical link between innovation and economic growth has been contemplated as early as Adam Smith (1776). Not only did he articulate the productivity gains from specialization through the division of labour as well as from technological improvements to capital equipment and processes, he even recognized an early version of technology transfer from suppliers to users and the role of a distinct R&D function operating in the economy.

Kuczmarski [5] sees innovation broadly and as a way of "thinking focused beyond the present into the future", as "intangible and intuitive, … a mindset". The concept of newness associated with innovation is referred to by other authors such as Slappendel [6] who sees the perception of newness as "essential to the concept of innovation as it serves to differentiate innovation from change", while

Hage [7] highlighted the distinction between "radical" and "incremental" innovations. The European Green Paper on Innovation [8] takes a broader stance and defines it as "the successful production, assimilation and exploitation of novelty in the economic and social spheres". Nonaka and Takeuchi [9] and many others speak of innovation as the key element of business success and draw parallels between knowledge and innovation. Others such as Walsh [10] add to the perspectives on innovation, stating that an innovation is only accomplished after the first 'commercial transaction'. Thus, an innovation is only achieved when having proved value and relevance in the market place.

Today, innovation is facing new challenges. Its own dynamism has produced a world that requires in many ways a rethinking of innovation itself. In the corporate sector, the determinants of innovation performance have changed in a globalized knowledge-based economy, partly as a result of recent developments in information and communication technologies. Strategies like market capitalization, mergers and acquisitions and just-in-time delivery, have to be revised in the light of the Internet, virtual organizations, online shopping and many other new approaches to creating value for customers.

7.4 The Learning Economy

Innovation is crucial to the competitiveness of all economies, and knowledge and learning is crucial to innovation. As we have seen in the previous chapter and in public debate and policy making, knowledge is increasingly presented as the crucial factor in the development of both society and the economy. In a growing number of publications from the European Commission and the OECD, it is emphasized that we currently operate in a "knowledge-based economy". For several reasons many prefer the term "the learning economy" in characterizing the current phase of socio-economic development [11]. The term illustrates the dynamic aspects of the economy, how companies needs to innovate and continuously adapt to and benefit from a changing business environment. The companies need to learn to survive and flourish in the same way as people and employees continuously have to learn to be attractive for companies. The most important trend shift is not that knowledge is becoming more important, but that it is becoming obsolete more rapidly than before, so that firms and employees constantly have to learn and acquire new competencies.

Knowledge and learning are key aspects of the OECD's suggested strategy for maintaining competitiveness. The OECD Jobs Study in 1994 highlighted the importance of skills and learning with respect to developing effective policy recommendations for the modern economy [12, p. 37]:

> Life-long learning must become a central element in a high-skills, high-wage jobs strategy.

Several implications were discussed in the OECD report. For example, recommendations include more active labor market policies, significant increases in the resources devoted to education and job-related skill development, and appropriate macroeconomic policy to reduce unemployment and promote economic growth. This involves different types of knowledge of which the less formalized—'learn through experience', are often just as important as the formalized—'learn through exposure to teaching'. A company being part of the learning economy means that it is able to not only adapt to changes in the business environment, but also in many cases that is a driver for change, being proactive through innovations and using new knowledge. The companies need to have a structure that enables flexibility and knowledge creation and knowledge transfer.

In a learning economy education and training in all parts of the organization is important. The employees increasingly need to have specialized and formalized education and skills, but also the flexibility and ability to combine knowledge and skills. Value is less and less created vertically through command and control, as in the classic 'teacher instructs student' relationship, but horizontally, by whom you connect and work with, whether online or in person. In other words, we are seeing a shift from a situation where knowledge is stored up but not exploited, so that it depreciates rapidly, to a situation of flows, where knowledge is energized and enriched through communication and constant collaboration. This approach to learning will become the norm. Barriers will continue to fall as skilled people appreciate, and build on, different values, beliefs and cultures.

In the efforts for continuous improvement and development companies need employees at shop floor level, who not only are involved but also in many cases, are the most important resources for these processes. This is reflected in *kaizen* and other basic principles within lean manufacturing. Getting there is not evident, it requires a business culture and incentives where people are eager for knowledge and skills, and where this knowledge and skills are applied in continuous learning and improvement within the company.

7.5 Social and Environmental Renewal

Sustainable manufacturing is commonly described as consisting of social, environmental and economic perspectives. This has also been focused on under the strategy of the Lisbon Agenda. The idea is that a stronger economy will create employment in the EU, alongside inclusive social and environmental policies, which would themselves drive economic growth even further. The main fields in the Lisbon Agenda are economic, social, and environmental renewal and sustainability. As previously discussed the way to get there is through:

- Innovation as the driver for economic change
- The "learning economy"
- Social and environmental renewal.

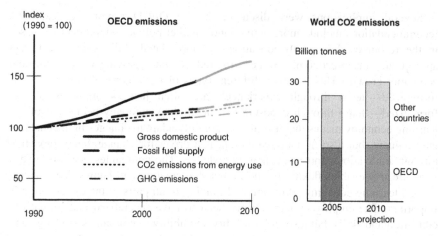

Fig. 7.3 Indicators for climate change [15]

In his first speech in November 2009 after being appointed president of the European Council, Herman Van Rompuy said [13] "Europe needed to double its economic growth rates: not only so that we can play our part in the world but above all to safeguard the achievements of our European way of life."

European countries together with other industrial countries still have challenges in making the step necessary to move into a radically changed social and environmental policy. This is not only due to difficulties in implementing the strategies or tasks required, but also a matter of discrepancies in what should be the policy and goals of change. This means that it is not evident where to find the balance between the three corners of the 'sustainability triangle'.

The Study Group Europe [14] states that only a fundamentally new approach to coordination of economic and employment policies can make the planned EU strategies a success for the citizens of Europe. According to this study group a new ideal-model beyond the mantra of absolute competitiveness must be drawn from the central challenges arising from the international economic and financial crisis, climate change, demographic developments and increasing social disparities. This requires a sustainable prosperity strategy and enhancement of various instruments of change.

There is obviously no difficulty in documenting the need for strategies for increased attention on the environment and sustainability in manufacturing. Even through improvements within certain fields and regions have been documented the overall picture shows us that knowledge creation innovations and radical measures are required. The OECD environmental indicators [15] could help us to not only get the overall picture but also to make sustainable strategies at company level, for example related to location, markets and customer preferences (Fig. 7.3).

The main concerns are on effects of increasing atmospheric greenhouse gas (GHG) concentrations on global temperatures and the earth's climate, and consequences for ecosystems, human settlements, agriculture and other socio-economic activities. This is because CO_2 and other GHG emissions are still growing despite some progress

Fig. 7.4 Indicators for waste generation [15]

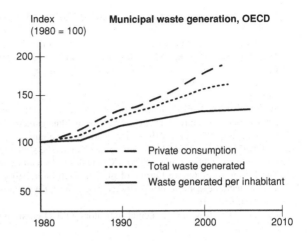

being achieved in decoupling CO_2 emissions from GDP growth. The main challenges are to limit emissions and to stabilize the concentration of GHG in the atmosphere at a level that would limit dangerous anthropogenic interference with the climate system. While a number of OECD countries have decoupled their CO_2 and other GHG emissions from GDP growth, most countries have not succeeded in meeting their own national commitments. Their emissions continued to increase throughout the 1990s, despite gains in energy efficiency. Overall, since 1980, CO_2 emissions from energy use have grown more slowly in the OECD countries than they have world-wide.

Similar trends are reflected in the indicators for waste reduction. The main challenge is to strengthen measures for waste minimization, especially for waste prevention and recycling, and to move further towards life cycle management of products and extended producer responsibility. From Fig. 7.4 we see that waste generation has increased considerably, but less than the private consumption in the OECD countries [14].

Even if it has proved too difficult to agree upon global agreements, environmental issues have been important in policymaking at national, but also regional levels. One important field where environmental aspects have been important has been within government funding of research and development. In the EU we find many programs within environment research fields, but environmental aspects are also increasingly highlighted in research programs for industrial innovations and increased competitiveness.

References

1. UNESCO (2010) Global education digest 2010
2. http://www.oecd.org/document/. Accessed Jan 2011
3. http://europa.eu/european_council/conclusions/index_en.htm. Accessed Nov 2006
4. http://www.oecd.org/document/10/0,3746,en_2649_33723_40898954_1_1_1_1,00.html, 03 July 2011. Accessed July 2011

5. Kuczmarski T (1996) What is innovation? The art of welcoming risk. J Consum Mark 13(5):7–11
6. Slappendal C (1996) Perspectives on innovation in organizations. Organ Stud 17(1):107–129
7. Hage JT (1980) Theories of organizations: form, process, and transformation. Wiley, New York
8. European Commission (1995) The European green paper on innovation
9. Nonaka I, Takeuchi H (1995) The knowledge creating company. Oxford University Press, New York
10. Walsh V (2002) Brief history of economic thought; demand, markets, and selection environments. Lecture. Manchester School of Management, Manchester
11. Lundvall BÅ, Johnson B (1994) The learning economy. J Ind Stud 1(2):23–42
12. OECD (1994) The OECD jobs study; facts, analysis, strategies
13. http://www.youtube.com/watch?v=hXWeOa-FuyM&feature=related. Accessed Oct 2011
14. The Berlin Study Group Europe of the Friedrich-Ebert-Stiftung (2010) Paving the way for a sustainable european prosperity strategy. http://library.fes.de/pdf-files/id/ipa/07053.pdf. Accessed Oct 2011
15. OECD (2008) Key environmental indicators. OECD environment directorate. Paris, France

Chapter 8
Research Roadmaps in Manufacturing

Abstract The contribution of manufacturing to the momentum to change and improvement in modern civilization has its basis in innovations and improvements that to a large extent build on research activities. According to OECD manufacturing accounts for the bulk of business R&D. 80% of the EU private sector Research and Technology Development expenditure is spent within manufacturing. Much research work has been devoted to improving manufactured product quality and manufacturing process efficiency for many decades. However, new research areas in manufacturing are emerging and we also see that new research schools are now entering the manufacturing field, such as social science, strategy development, and increasingly innovation, knowledge and learning. Over the last 40–50 years manufacturing industry has undergone a significant change, new manufacturing paradigms have appeared with new focus areas. It is possible to recognize four generations of research focus areas: machine focus, factory focus, supply chain focus and now, life cycle focus.

8.1 Research Resources in Manufacturing

Manufacturing contributes significantly to modern civilization and creates momentum that drives today's economy. This momentum has its basis in innovations and improvements that to a large extent stems from research activities. Manufacturing companies need in one way or another to take advantage from these knowledge creation processes to over time survive in the increasingly changing and complex market place.

In Fig. 8.1 we see that at OECD level, manufacturing has accounted for the bulk of business R&D [1]. In Europe, there are close to 540,000 manufacturing researchers. 72% of the EU researchers in the Business Enterprise Sector work in manufacturing, engineering and technology, at the same time researchers

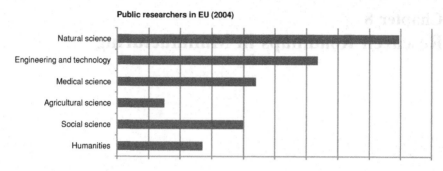

Fig. 8.1 Public researchers in EU-25 in full time equivalent (2004) [1]

constitute 21% of total public researchers in EU. Finally, 80% of the EU private sector Research and Technology Development expenditure is spent within manufacturing.

Much research work has been devoted to improving product quality and manufacturing process efficiency. This remain basic issues on the research agenda, and thanks to recent advances in computer and network technologies, sensors, control systems and manufacturing machines, manufacturing research has progressed to a new level. However, new research areas in manufacturing are emerging to address problems encountered in changing manufacturing environments that the increased practice of for example globalization and outsourcing represent. We also see that other research schools have entered into the manufacturing field, such as social science, business strategy, and increasingly innovation, knowledge and learning.

8.2 Shift in the Manufacturing Research Agenda

Over the last 40–50 years manufacturing industries has undergone a significant change, new manufacturing paradigms have appeared with new focus areas. These changes are clearly demonstrated in the shift of the manufacturing research agenda. It is possible to recognize four generations of research focus areas [2]:

- Machine focus
- Factory focus
- Supply chain focus
- Life cycle focus

In the machine focus, research activities are targeted at a single process element, for example around the cutting process in a single machine tool. Improved cutting data, cutting conditions, cutting materials, tools and capability of the machine tool are all typical examples of problem areas that have been improved significantly during this generation of research.

The second generation changed the focus from the single process element to the whole factory. Research focus is on industrial automation, robotics, Computer Numerical Control (CNC), flexible manufacturing, operations and process planning, and in the end on Computer-aided design/Computer-aided manufacturing (CAD/CAM), Computer integrated manufacturing (CIM), and intelligent manufacturing where integrated computers control the entire production process.

The third generation shifted the focus to the whole supply chain. In addition to manufacturing technology, the research now has a dimension focused on management and economics. One of the main enablers of this research is the availability of digital solutions for planning, control, management and business operation. The 'extended enterprise' is a topic that has received much attention in this respect [3].

The fourth generation maintains the focus on the supply chain, but extends this into a full life cycle of the product. The focus shifts to a certain extent from production to the product. The concept of extended products has been launched covering the whole life cycle from delivered product, through use and maintenance, and on to disassembly and recycling. The life cycle approach has been adopted or maybe even initiated by the increased focus on corporate social responsibility and sustainable manufacturing. The life cycle approach covers a broad set of research areas but an important part of it has been to develop methodologies and technologies to track products and resources through the entire life cycle, and to create knowledge and decision support from it.

The framework developed by the GEM-project[1] identified seven core knowledge areas within any new manufacturing curriculum and all of which reflect the current and future needs of the manufacturing industry. These areas defined in 2004 are still basis for important research projects in manufacturing. Table 8.1 shows an overview of these knowledge areas. Each area focuses on aspects of manufacturing that were considered future oriented and promoting a paradigm shift from traditional manufacturing to digital business. Each area comprises courses and individual subjects (equivalent to an individual lecture). Each course is described and structured into a number of subjects and with a recommendation for their delivery.

8.3 The IMS2020 Roadmap

IMS2020 was launched in January 2009 as a 24 month project co-funded by the European Commission and an international consortium of 15 companies, universities and research centres from countries in Europe, US and Asia.

---

[1] Global Education in Manufacturing (GEM) was an IMS project that was endorsed in January 2002 and was finished in November 2004. The European part was sponsored by the European Union (FP5 Research Program).

Table 8.1 GEM knowledge areas (2004) [2]

Knowledge area	Description
A Development of extended products	The development of a combination of a physical product and associated services/enhancements that improve marketability
B Digital business along the supply chain	Information on how a business can use e-commerce and related technologies and processes to develop, expand or enhance its business activities along the facilities and functions involved in producing and delivering a product or service
C End of life planning and operation	Techniques on how to develop methodologies and tools to support the end-of-life routing/processing decision based on economic, environmental and societal criteria
D Business operation and competitive strategy	Explanation of how organizations function and interact with competitors and their market place, and deliver performance over time
E Intelligent manufacturing processes	Elaboration of techniques applicable for handling complex production working in an uncertain, changing environment, with special emphasis on artificial intelligence and machine learning approaches
F Intelligent manufacturing systems design	Tools on how to model the skills and knowledge of manufacturing experts so that intelligent equipment and machines can produce products with little or no human intervention
G Enterprise and product modelling and simulation	Information on how to develop and use computational representations of the structure, activities, processes, information, resources, people, behaviour, goals and constraints of a business or a product

The aim of IMS2020 has been described as predicting the challenges to be met if the manufacturing industry is to undergo the "deep industrial transformation" which experts say is needed to meet the environmental, social and economic challenges of tomorrow. The IMS2020 research roadmaps describe to meet the IMS 2020 vision [4]:

- Rapid and adaptive user-centered manufacturing which leads to customized and 'eternal' life cycle solutions
- Highly flexible and self-organizing value chains which enable different ways of organizing production systems, including related infrastructures, and reduce the time between engaging with end users and delivering a solution
- Sustainable manufacturing possible due to cultural change of individuals and corporations supported by the enforcement of rules and a proper regulatory framework co-designed between governments, industries and societies.

The IMS2020 roadmaps describe a number of research topics and supporting actions which need to be fostered through international cooperation. These are critical research topics which, when implemented, will allow the achievement of the defined IMS2020 vision and thus the shaping of manufacturing systems by the year 2020 and beyond.

There are five roadmaps, each covering a specific key area. KAT 1, 2 and 3 cover the technology elements and comprise [4]:

- Sustainable manufacturing (KAT 1)
- Energy efficient manufacturing (KAT 2)
- Key technologies (KAT 3).

Standards (KAT 4) and education (KAT 5) are both associated with all the three other areas. Standards are of course valid for sustainable manufacturing, energy efficient manufacturing and key technologies. The same is true for education.

The IMS2020 roadmap for sustainable manufacturing is aimed at improving the sustainability of the technologies, the products and production systems as well as the businesses behind them. The research topics have been clustered into five research actions [4]:

- Technologies for sustainability
- Scarce resources management
- Sustainable lifecycle of products and production systems
- Sustainable product and production
- Sustainable businesses.

Development of the KAT5 (education) roadmap has been done in close cooperation with the other KAT Roadmaps. Nine research topics have been identified [5]:

- Teaching factories
- Cross sectoral education
- Communities of practice
- From tacit to explicit knowledge
- Innovation
- Benchmarking
- Serious games
- Personalized and ubiquitous learning
- Accelerated learning.

In addition there are six research topics for innovation identified [5]:

- Global brain
- Learn to innovate
- Go green
- Living labs
- Risk management
- Sustainable new apprenticeships.

As we see the IMS2020 project not only point out key areas for future research and development within manufacturing, but also the means and enablers for reaching those goals and the overall visions. Means and enablers for knowledge creation and learning are added to the more traditional institutional education systems. They are consequences of rapid change of competence needs in industries and an increased average age of employees which also require new education and training systems. Development of new training methods is by IMS2020 viewed as a priority for education investment in most advanced countries and has become a proper research field within manufacturing.

References

1. UNESCO "Global Education Digest 2010"
2. O'Sullivan D, Rolstadås A, Filos E (2009) Global education in manufacturing strategy. J Intell Manuf 22(5):663–674
3. Jagdev HS, Browne J (1998) The extended enterprise—a context for manufacturing. production. Planning Control 9(3):216–229. Taylor and Francis
4. Taisch M, Cassina J, Cammarino B, Terzi S, Duque N, Cannata A, Urgo M, GarettiM, Centrone D, Ibarbia JA, Kiritsis D, Matsokis A, Rolstadas A, Moseng B, Oliveira M, Osteras T, Vodicka M, Bunse K, Cagnin C, Konnola T, Oedekoven D, Bauhoff F, Trebels J, Hirsch T, Kleinert A, Carpanzano E, Paci A, Fornasiero R, Chiacchio M, Rusinà F, Checcozzo R, Pirlet A, Brülhart M, Ernst F (2010) Action roadmap on key areas 1, 2 and 3. IMS2020 project report, POLIMI, Milano
5. Rolstadas A, Moseng B, Vigtil A, Osteras T, Fradinho M, Carpanzano E, Bromdi C (2010) Action roadmap on key area 5. IMS2020 project report, Trondheim, Norway

Chapter 9
How Well Are We Doing?

Abstract Performance is closely linked to competitiveness which again is closely linked to productivity and is therefore of significant interest for manufacturing enterprises competing in global markets. Companies have to find ways to measure and improve how well they are performing in innovation, knowledge creation and learning, but also the other strategic areas that traditionally have been measured such as productivity, cost and quality. Performance measurement is important, not only to recognize and understand the extent to which a company reaches its goals and objectives, but also to guide action and strategic direction. To measure performance there has to be an agreement upon what performance means. For the intangible aspects of change, we must rely on proxies or indirect measures i.e. indicators that often only capture a fraction of what we want to measure. Measuring knowledge on a more strategic level could be done by focusing on the capabilities they are supposed to contribute to. Relevant frameworks could include: Performance Prism, Balanced Scorecard, Quality Function Deployment (QFD), and Value Creation Maps.

9.1 Measuring Performance in Companies

In earlier chapters we have seen that countries and regions are performing differently when it comes to competitiveness. We have derived several indicators for competitiveness and see that innovation, knowledge creation and learning are increasingly important. This is in particular also the fact for manufacturing companies in the global economy. National or regional performance would normally be an aggregation of what the companies are doing. However, when companies or industries are measuring performance we are moving into a detailed level that is much higher, but which also makes it possible for them to develop strategies and concrete activities for improvement and development. Performance is closely linked to competitiveness which again is closely linked to productivity and is therefore of significant interest for enterprises competing in the global markets.

Performance measurement is a key for successful implementation of strategies. Since knowledge is an increasingly important dimension of strategy, companies need to find measures for these aspects. Companies have to find ways to measure and improve how well they are performing in innovation, knowledge creation and learning, but also the other strategic areas that traditionally have been measured such as productivity, cost and quality. Performance measurement is important, not only to recognize and understand the extent to which a company reaches its goals and objectives, but also to guide action and strategic direction. People and organizations have a tendency to act according what is measured, especially when they are rewarded according to such measurements [1].

Measuring of performance should be done from different viewpoints. There are normally three different dimensions:

- *Effectiveness*—to which extent are customer needs satisfied
- *Efficiency*—to which extent are the total resources in the company used in an effective and economic way
- *Ability to change*—to which extent are the company prepared to handle changes in surrounding conditions (strategic awareness)?

When establishing a methodology for productivity measurement it is important to have in mind all three dimensions. In addition it is important to cover all key stakeholders as they will significantly influence performance and productivity.

The classic definition of productivity is produced goods per unit of production. This definition indicates a focus on the value added during the physical production process. The physical production process is well defined in most enterprises, and input and output can easily be counted, measured or calculated. Internal services, however, are not so easy to describe, and even more difficult to measure. One of the first approaches to performance measurement was published by Sink [2] and Tuttle [3]. The model claims that the performance of an organizational system is a complex interrelationship between the following seven criteria:

- Effectiveness
- Efficiency
- Quality
- Productivity
- Quality of work life
- Innovation
- Profitability/budgetability.

Effectiveness means doing the right things, at the right time, with the right quality. Defining the criterion as a ratio, effectiveness can be defined as 'Actual Output/Expected Output'. Figure 9.1 illustrates this. Efficiency means doing the right things and is a term connected to resource utilization and can be defined as the ratio 'Resources Expected to Be Consumed/Resources Actually Consumed',

Fig. 9.1 Operational definition of effectiveness

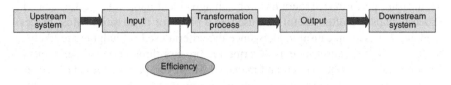

Fig. 9.2 Operational definition of efficiency

Fig. 9.3 Operational definition of productivity

as shown in Fig. 9.2. Productivity is the traditional ratio between output and input as illustrated in Fig. 9.3.

When Sink and Tuttle introduced their seven performance criteria, mainly focusing on productivity and quality, it was an implicit description of the focus in manufacturing strategies. In a changing and increasingly complex strategic context performance measurements need to be aligned. Defining measures, and a system for collecting and analyzing them, is particularly difficult for the intangible aspects related to infrastructure, such as knowledge and coordination issues. Customer orientation, flexibility, partnership, decentralization, continuous improvement, knowledge and innovation, are some of the aspects that have received increased attention in lean manufacturing and other more recent paradigms. To measure how the companies perform they also have to capture intangible aspects, but that may be difficult since we then have to accept measures that may not be accurate.

Indicators and qualitative measures might be difficult to derive. Qualitative measures can be accurate, but also resource demanding and difficult to implement as an integrated part of operations. Since intangible aspects of manufacturing operations often have to be measured through indicators we must be aware that they often only capture a fraction of what we want to measure, and have to be treated accordingly. Adding intangible aspects to the performance measurement system increase the risk of having too many metrics. Too many metrics can increase the administrative burden and can be counterproductive in achieving improvements. We believe that this process will be better off if performance measurement systems are developed with a broad involvement of employees and partners.

9.2 Measuring Performance According to Strategy

To measure performance there has to be an agreement upon what performance is. Process maps are often useful and used as a basis for discussions. Most process mapping is based on causal models since the criteria used for judging are likely to focus on results. Causal models describe the process through which performance (future results) will be created and managed. However, past performance alone is not necessarily a good predictor of future performance [4]. Marr [5] has presented the 'bundle value creation map', corresponding to a dynamic view and interdependencies. Lebas and Euske define performance as:

> .. the sum of all processes that will lead managers to taking appropriate actions in the present that will lead to measured value tomorrow. [4, p. 68]

Having agreed upon what performance means, the next and difficult task is measuring it. From the above definition of performance it is logical that a key measure of corporate performance is 'added value'. This could be defined as the difference between the (comprehensively accounted) value of a firm's output and the (comprehensively accounted) cost of the company's input [6]. There is often a very ambiguous use of measurement, with too many metrics that no one uses, and others that are not measuring what they are supposed to measure [5]. Austin and Gittell [7] present what have traditionally been considered basic principles of performance measurement:

1. Performance should be clearly defined: if not in advance, then in terms of criteria that can be agreed after the fact
2. Performance should be accurately measured: as work is performed, performance should be measured in a way that conveys the maximum amount of information
3. Reward should be contingent upon measured performance.

Austin and Gittell's principles are still relevant for some purposes and types of measurement, but clearly imply risks such as just measuring what can be counted (principle no. 1), data overload (principle no. 2), or just rewarding behavior producing quantifiable outcomes (principle no. 3). The principles become even more questionable when the often intangible aspects of strategies are to be measured.

The four classic measures of manufacturing are: cost, time, quality, and flexibility [8], but we normally have to employ a balanced set of measures of tangible and intangible aspects to obtain the best picture of performance and be able to locate areas for improvement. There are several dimensions of measures such as [1]:

Table 9.1 Bohn's stages of knowledge

Stage	Name	Comment	Typical form of measures
1	Complete ignorance	–	Nowhere
2	Awareness	Pure art	Tacit
3	Measure	Pre-technical	Written
4	Control of the mean	Scientific method feasible	Written and embodied in hardware
5	Process capability	Local recipe	Hardware and operating manual
6	Process	Trade-offs to reduce costs	Empirical equations (numerical)
7	Know why	Science	Scientific formulas and algorithms
8	Complete knowledge	Nirvana	–

- Hard–soft
- Financial–non-financial
- Result–process
- Objective/quantitative–subjective/qualitative

Measures can also often be defined by their purpose for example diagnostic or to motivate for a certain change in attitude.

For the intangible aspects, we must rely on proxies or indirect measures—indicators that often only capture a fraction of what we want to measure and have to be treated as such. One way of balancing measurements is through triangulation, whereby a company collects data from different data sources, uses different methodologies or uses different people in data collection.

Measuring knowledge on a more strategic level can be done by focusing on the capabilities they are supposed to contribute to. Some measures are presented in the literature, such as: product customization ability, volume flexibility, mix flexibility, and time to market [9], or 'qualifiers' and 'order winners' [10]. Kaplan and Norton [11] state that measures related to the response time, quality, and price of customer-based processes are typically incorporated in virtually all value propositions. However, capabilities may also be difficult to recognize and categorize, as they often affect several competitive priorities at the same time (i.e., price, flexibility, delivery, quality, and service), and are normally unique to each company [12, 13].

Table 9.1 lists Bohn's [14] stages for measuring "technical knowledge", ranging from complete ignorance to complete understanding. Bohn's measures are related to knowledge about a particular input variable for a particular process output.

Performance measurement should be an integrated part of strategy development as it provides analysts and decision makers with data and information. Marr [5, p. 4] defines strategic performance measurement as:

> The organizational approach to define, assess, implement, and continuously refine organizational strategy. It encompasses methodologies, frameworks and indicators that help organizations in the formulation of their strategy and enable employees to gain strategic insights which allow them to challenge strategic assumptions, refine strategic thinking, and inform strategic decision-making and learning.

There are many approaches and methods that can be used to guide companies towards key measures and enable them to work in a more structured way on strategic performance measurement. There are several evaluations of frameworks (see for example, Bourne et al. [15] and Busi [16]). Even if ICT today enables data collection and measures in much broader business areas and processes than before there are still a need for performance data that are hard to capture, especially the increasingly important intangible aspects. The following section describes how strategic and intangible aspects can be measured through different frameworks.

9.3 Frameworks for Strategic Performance Measurement

Strategies are developed as interplays between different stakeholders' interests. Atkinson et al. [17] define two groups of stakeholders: "environmental" (customers, owners, and the community) and "process" (employees and suppliers). Neely and Adams, [18] present five distinct, but logically interlinked, questions and perspectives on performance:

- Who are the key stakeholders and what do they want and need
- What strategies do we have to put in place to satisfy their wants and needs
- What are the critical processes required to execute these strategies
- What capabilities do we need to operate and enhance these processes
- What contributions will this require from our stakeholders?

Neely and Adams' *Performance Prism* in Fig. 9.4 shows the complexity of performance measurement and management, and focuses on aligning processes with strategies and capabilities. One dimensional, traditional frameworks pick up elements of this complexity, but to understand it in its entirety it is necessary to view it from the multiple and interlinked perspectives. An advantage of the stakeholder approach is the understanding of different stakeholders having different views on what should be measured [19]. The intangible aspects, and their subjective character, require specific attention as they are not always clearly expressed by the stakeholders.

The Balanced Scorecard [11] (Fig. 9.5) has become a popular strategic performance approach within many industries. The framework is used to translate an organization's vision and strategy into a comprehensive set of performance perspectives and measures. The approach focuses on a few key aspects, breaking down strategic measures to lower levels so that operating managers and employees are able to see the strategic implications of their work. It allows measuring present performance, but also how the organization is positioned for the future through learning and growth.

Methods that are not specifically designed as frameworks for strategic performance measurement can have links to strategic levels or a hierarchy of measures that indicate strategic impacts and thus be useful as frameworks for strategic performance

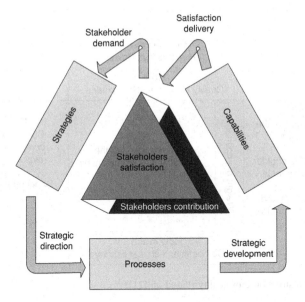

Fig. 9.4 The performance prism [18, 19]

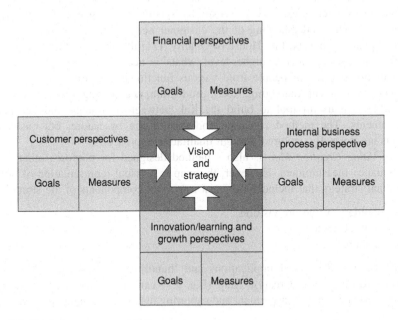

Fig. 9.5 The balance scorecard [11]

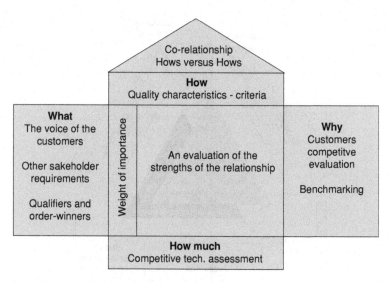

Fig. 9.6 QFD matrix/house of quality (based on [21])

measurement. *Quality Function Deployment* (*QFD*) is used to translate customer requirements to engineering specifications and setting targets [20].

The 'voice of the customer' is captured in a variety of ways, through direct discussion or interviews, surveys, customer specifications, warranty data, field reports, etc. This understanding of the customer needs is then summarized in a product planning matrix, the 'House of Quality' [21] illustrated in Fig. 9.6. These matrices are used to translate higher level 'whats' or needs into lower level 'hows'. QFD involves teams of people from various functional departments engaged in product development: marketing, R&D, production, engineering, product support, etc. QFD is a useful tool to build the link between a strategy and customer requirements. The method is causal and quantifying by nature, but customer aspects are normally captured through qualitative approaches.

The *Value Creation Map* is also a method designed for purposes other than performance measurement. However, the map is a visual representation of organizational strategy that includes the most important components of the strategy:

- Stakeholders' value proposition
- Core competencies
- Key resources

By revising the causal implications and 'bundling' the resources, Marr [5] shows how the Value Creation Map (Fig. 9.7) can serve as a framework for strategic performance measurement and capturing dynamic and intangible features of strategy.

Other types of process maps or ways of describing organizations might represent feasible frameworks for strategic performance measurement as long as they have links to overall objectives and stakeholders requirements.

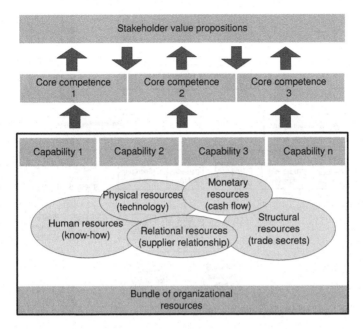

Fig. 9.7 Value creation Map [5]

9.4 Decomposition, the Logic Behind the Strategy

The choice of performance measurement framework, many of which have been described earlier, should be based on the purpose of the performance measurement system and aligned to strategy. If a company is not able to derive measures and build a performance measurement from its manufacturing strategy, it is an indication that the strategy that is not sufficiently concrete to be a guide for action. The approaches and frameworks presented are examples of performance measurement frameworks, but other frameworks and approaches could be used even if they have been developed for other purposes. Quality systems and different kinds of process descriptions are often used in performance measurement and may be relevant for different manufacturing strategies.

Measuring how well a company performs according to for example the 'lean strategy' is difficult to measure and requires a focus on intangible aspects. A structured and more detailed description is needed. Figure 9.8 shows an example of a model that can be applied to ease the task of identifying measurement areas and indicators for lean manufacturing [22].

It can be seen that there is still a probability that the intangible aspects of basic lean principles will not be captured, such as the 4-P model [23]:

- A *philosophy* of long-term thinking
- Eliminating waste in *processes*

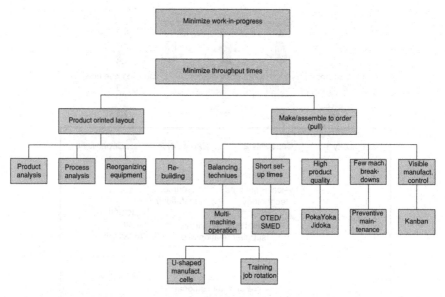

Fig. 9.8 The logic behind lean manufacturing [21]

- Respect *people and partners*, challenge and develop them
- *Problem solving*, continuous improvement and learning.

There is still a tendency in lean manufacturing to measure those things that are easy to measure and backward looking, even when the strategy is defined around lean principles. Aspects such as product quality, Single Minute Exchange of Dies (SMED), Just in Time (JIT), and time to market, are important and easy to quantify and measure. The more dynamic, intangible aspects, such as customer/supplier relations, learning and incremental innovation, and teamwork, are more difficult to quantify, and hence less easy to measure.

9.5 Performance Measurement Implementation

There are many ways to implement performance management in industry. All of them require a feasible system for performance measurement. Performance measurement can be done at different levels:

- Internal in a company
- Between selected companies
- At a national level (discussed in Chap. 6 and 7)
- At an international level (discussed in Chap. 6 and 7)

Internally within a company there can be agreed indicators to be monitored between departments or business units. Between companies indicators can be obtained by agreeing to exchange data. However, a more common way is to undertake a benchmark organized by a consultancy. The consulting company would then own a database against which a company can benchmark against making its own data anonymously accessible to others. At a national or international level, it would typically be done as a research task.

Such a research task was undertaken in Norway in the 1990s. The research program (called TOPP) consisted of four sub programs [24]:

- Analysing company productivity and competitiveness (self-audit, extended audit, benchmarking, etc.)
- Implementing projects for industrial productivity improvements (industrial projects, seminars, courses, industrial networks, etc.)
- Generating new knowledge (R&D projects analysing productivity data, etc.)
- Long term competence programme (education, courses, doctorate in engineering sciences, master's degree, etc.)

About 60 enterprises participated in the program. Performance was measured using four different approaches:

- Self-audit based on a questionnaire
- Extended audit performed by external experts
- Self-assessment (continuous improvement)
- Benchmarking (breakthrough)

The self-audit was based on a questionnaire answered by the companies. In the questionnaire the companies were asked to evaluate the different functions of the company on a scale from 1 to 7. This indicated the current status and where company expected to be in 2 years. In addition companies ranked the importance of the indicators as low, medium or high. The questionnaire was answered by selected individuals in the company, and by defined groups. Individual and group views was found to differ in some cases. The self-audit was based on a predefined functional breakdown of the company. This can be valid for a given industrial sector, but not necessarily across industrial sectors. The functions were split on primary functions and support functions. The primary and support functions for a manufacturing company are shown in company breakdown model in Fig. 9.9. Each function was broken down one level further for the questions.

The extended audit was performed by external experts doing in depth analysis of each company. The analysis was performed at two different levels:

- Company level using indicators and key factors with focus and viewpoint on the whole enterprise
- Company split up level using indicators and key factors with focus on parts and specified areas in the company

At the company level the following indicators were used (see Fig. 9.10):

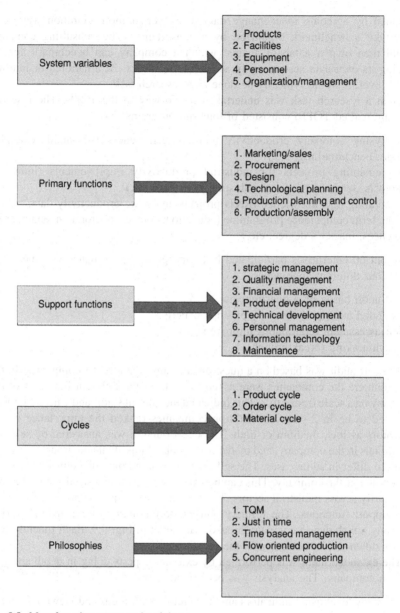

Fig. 9.9 Manufacturing company breakdown

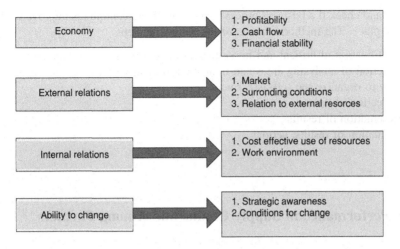

Fig. 9.10 Indicators at company level

- *Economy*—indicators and key factors which describes the economic conditions and potential. An important issue of this analysis is to evaluate the economic capacity for future investments and change
- *External relations*—indicators and key factors to measure the ability of the company to exploit and take advantage of surrounding conditions. This includes evaluation of customers, competitors, use of external resources, strategic alliances, etc.
- *Internal relations*—the ability to use existing internal resources (products, machines, personnel, etc.) in a time and cost effective way, and to take care of the internal milieu
- *Ability to change*—the ability to foresee and be prepared to meet new trends and quick changes in the environments.

At the company functional level, a functional breakdown was used:

- *Systems variables*—primary resources and conditions necessary to produce products
- *Functions*—functions in the product life cycle. The functions are divided in primary and support functions
- *Cycles*—to follow and analyze flows between functions. Flows could be of different types (material, information, etc.). The objective is to measure co-operation and infrastructure between functions
- *Philosophies*—to analyze overall production and management philosophies used in the company.

The self-assessment approach is a modular system to help the companies to build their own system for performance measurement. The objective is to follow trends and check if a business process is in control and continuous improvement is taking place. The method consists of the following steps

- Identification of critical and important business processes in the company
- Selection of analysing areas and indicators to measure the business processes
- How to organize the self-assessment
- Collection of data
- Presentation of results
- Evaluation of results, actions.

9.6 Performance in Supply Chains and Among Partners

A companies' performance is increasingly a question of interaction between different units and partners. Improvement in supply chains requires collaborative efforts and the ability and willingness to share data and knowledge is crucial.

Performance among partners in a supply chain is increasingly measured and structured through ICT, such as Enterprise Resource Planning (ERP) applications. These applications can become frameworks for performance measurement. There have also been developed performance management software applications aiming for implementation of frameworks such as Balanced Scorecard, including data integration, analysis and communication of real time management information [11].

Automated data gathering, enabled by the increased automation of production and processes, represents an opportunity for improved performance measurement. This has, for example, been one of the driving forces for adaptive manufacturing. New enabling technologies also facilitate reporting from, manufacturing satellites to the mother plant, and hence could support some kind of centralized decision-making characterizing mass manufacturing [24]. More integrated business processes in supply chains have also been supported by ICT enabling data collection and knowledge sharing.

In sustainable manufacturing, life cycle assessment (LCA) is a commonly used method for documenting environmental impact of products and processes. These methods rely on extensive frequent data collection from suppliers and other partners. Large databases aid the performance measurement but data quality can be questioned in many cases.

However, even if ICT continuously improves the potential for performance measurement between independent organizations there is still the risk that the aspects emphasized will be those which are quantifiable and easy to measure, transfer and analyze. The intangible aspects which are embedded in tacit knowledge and often related to continuous improvement are still difficult to measure.

References

1. Andersen B, Fagerhaug T (2002) Performance measurement explained. Designing and implementing your state-of-the-art system. ASQ Quality Press, Milwaukee
2. Sink DS (1985) Productivity management: planning, measurement, and evaluation, control, and improvement. Wiley, NY
3. Sink DS, Tuttle TC (1989) Planning and measurement in your organization of the future. Industrial Engineering and Management Press, Norcross
4. Lebas M, Euske K (2002) A conceptual and operational delineation of performance. In: Neely A (ed) Business performance measurement: theory and practice. Cambridge University Press, Cambridge, pp 65–79
5. Marr B (2006) Strategic performance management. Leveraging and measuring your intangible value drivers. Elsevier, Oxford
6. Kay J (1993) The structure of strategy. Bus Strategy Rev 4(2):17–37
7. Austin R, Gittell JH (2002) When it should not work but does: anomalies of high performance. In: Neely A (ed) Business performance measurement. Theory and practice. Cambridge University Press, Cambridge, pp 80–107 (Neely 2002)[92]
8. Blair M, Wallman S (2001) Unseen wealth. Brookings institution, Washington
9. Laugen BT, Acur NA, Boer H, Frick J (2005) Best manufacturing practices. What do the best-performing companies do? Int J Oper Prod Manage 25(2):131–150
10. Hill T (1983) Manufacturing's strategic role. J Oper Res Soc 34(9):853–860
11. Kaplan R, Norton D (1996) The balanced scorecard. Harvard Business School, Boston
12. Wheelwright SC, Bowen HK (1996) The challenge of manufacturing advantage. Prod Oper Manage 5(1):59–77
13. Pandza K, Polajnar A, Buchmeister B, Thorpe R (2003) Evolutionary perspectives on the capability accumulation process. Int J Oper Prod Manage 23(8):822–849
14. Bohn RE (1994) Measuring and managing technological knowledge. Sloan Manage Rev 36(2):61–73
15. Bourne M, Neely A, Mills J, Platts K (2003) Implementing performance measurement systems: a literature review. Int J Bus Perform Manage 5(1):1–24
16. Busi M (2005) An integrated framework for collaborative enterprise performance management. PhD thesis Trondheim, NTNU, p 48
17. Atkinson AA, Waterhouse JH, Wells RB (1997) A stakeholder approach to strategic performance measurement. Sloan Manage Rev 38(23):25–37 (Spring)
18. Neely A, Adams C (2001) The performance prism perspective. J Cost Manage 15(1):7–15
19. Bredrup H (1995) Performance evaluation. In: Rolstadås A (ed) Performance management: a business process benchmarking approach. Chapman & Hall, London, pp 191–198
20. Mizuno S, Akao Y (1978) Quality function deployment. Nikkagiren-Syuppan, Tokyo
21. Hauser JR, Clausing D (1988) The house of quality. Harvard Business Review, pp 63–73 (May-June)
22. Skorstad EJ (1999) Produksjonsformer i det tyvende århundre. Organisering, arbeidsvilkår og produktivitet ("Production in the 20th century. Organization, work conditions and productivity"). AD Notam Gyldendal, Oslo
23. Liker J (2004) The Toyota way: 14 managament principles from the world's greatest manufacturer. McGraw-Hill, NY
24. Andersen B, Printz Moe E, Moseng B, Rolstadås A (1996) Produktivitet og konkurranseevne i norske bedrifter. AD Notam, Oslo (in Norwegian)

Part III
Outsourcing: Strategic Opportunities

This part of the book describes the types of choices involved in creating manufacturing strategy and that include decisions around both structure and infrastructure. The role of the different manufacturing units along the supply chain and in the extended enterprise is a major strategic decision i.e., what activities are organized within and outside the specific company. This part describes how the outsourcing decisions have major implications for how to develop manufacturing capabilities.

Chapter 10
Manufacturing Strategies, Created Through Decisions

Abstract Manufacturing strategies essentially entail decisions about structure and infrastructure. Manufacturing structures deal with the vertical integration of operations, facilities and locations, capacity, and process technology. With increased focus on flexibility and customization, manufacturing strategies must also emphasize coordination and other infrastructure decisions. Infrastructural strategies include organizational and human aspects, sourcing and supply chain management practices, quality management systems, and other less tangible aspects. Infrastructure is developed over time through persistent day-to-day practice, top management commitment, and cross-functional efforts to create capabilities that support and leverage the firm's structure. How a company positions itself in the supply chain is an example of how structural decisions impact on knowledge creation and to some extent it defines what the company sees as its core activity. The core activity directs innovations which are incremental and more radical, and consequently it also directs knowledge creation.

10.1 Structure and Infrastructure, Hard and Soft Elements of the Strategy

Manufacturing strategies entail decisions around structure and infrastructure [1, 2], decisions that are assumed to be implemented to make a strategy real. Manufacturing strategies have traditionally focused on topics around fulfilling orders and adjusting capacity to meet the needs defined by markets [3]. As a consequence, manufacturing strategies have mainly emphasized decisions around manufacturing structures which deal with the vertical integration of operations, facilities and locations, capacity, and process technology [4].

The difficulty of dealing exclusively with these decisions has worsened due to increased competitive pressure, the acceleration of technological change, and the distribution of knowledge across various organizations and geographical markets along the supply chain. As a consequence of this there has been a shift in focus towards infrastructure

A. Rolstadås et al., *Manufacturing Outsourcing*,
DOI: 10.1007/978-1-4471-2954-7_10, © Springer-Verlag London 2012

decisions [5]. With increased focus on flexibility and customization, manufacturing strategies must also emphasize coordination and other infrastructural decisions. Infrastructure includes organizational and human aspects of manufacturing, sourcing and supply chain management, quality management, and other less tangible aspects [6]. According to Beckman and Rosenfield [4, p. 30] "Infrastructure is developed over time through persistent day-to-day practice, top management commitment, and cross-functional efforts to create capabilities that support and leverage the firm's structure".

If we are referring to resources versus capabilities, structural decisions will normally be related to resource issues and infrastructural decisions will be related to capabilities. Knowledge creation and knowledge transfer are also regarded as infrastructural issues. However, structural decisions such as location of facilities represent important prerequisites for knowledge creation and transfer. It is the effective integration and synthesis of structural and infrastructural decisions that create long term operational excellence [6].

10.2 Structure, The Physical Manifestation of a Strategy

Chandler [7] describes how structural decisions are 'hard' and relate to 'bricks and mortar' issues such as equipment and technology decisions that tend to support the resource deployment in the strategy. According to this categorization, ICT and other physical knowledge supporting technology are regarded as a part of structural decisions. Other researchers, such as Beckman and Rosenfield [4] make a distinction between process technology and information technology, where information technology is defined as infrastructure. However, in advanced manufacturing, for example adaptive manufacturing, it might be difficult to separate these two elements of technology. This indicates that it could be difficult to make distinctions between structural and infrastructural decisions.

Beckman and Rosenfield [4] categorize the following decisions as part of the structural element of manufacturing strategy:

- *Vertical integration*—how much of the supply chain should the company own? Should the company integrate upstream or downstream? This also reflects issues such as reliability, transaction costs, and risks related to quality and property. Opportunities for capability development and exploration of opportunities are also relevant issues
- *Process technology*—companies' investments in the technologies for transforming materials and/or information into products/service
- *Capacity*—how to balance fluctuations in demand and to how to respond to growth opportunities: How and when should capacity be provided? Should the company make use of contractors
- *Facilities*—questions related to location and the role of the different units. These questions are also concerned with size and number of facilities needed to meet the strategic objectives.

Choe et al. [8] discuss how structural decisions regarding process typically take an innovation focus, and consequently involve knowledge management. They define a manufacturing structure with a high level of process complexity and a low level of product complexity as a "process-innovative structure", as this emphasizes process innovation. A structure with a low level of process complexity and a high level of product complexity is termed a "product-innovative structure", as success will depend upon innovations related to products. They define an "integrative structure" as a strategy related to high levels of both product and process complexity, where innovations relating to product and processes need to be integrated. Consequently, Choe et al. also add a further definition, namely a "non-innovative structure", which is one that does not emphasize either process or product innovation.

Through its structural decisions, a company implicitly constrains its flexibility to alter its competitive priorities in the future. The decisions it makes effectively lock it into certain modes of behavior, making path dependencies for knowledge creation and knowledge transfer [5]. Structural decisions not only define what kind of knowledge is needed (process choice), but also who needs it and where to find it (vertical integration and facilities), and thus relate it to the transfer of knowledge.

In each of Beckman and Rosenfield's [4] categories, making decisions means striving to ensure that the decisions are mutually supportive and also consistent with each other and useful in an overall strategic context. What seems right today is not necessarily so tomorrow. In a business environment and competitive situation that changes rapidly we need 'room to maneuver'. When changing one of the structures or infrastructure elements, a company needs to consider how that change might affect its ability to make future changes to its priorities. From their study of 170 US manufacturing companies, Choe et al. [8, p. 409] conclude that "manufacturing companies can significantly improve their business performance by attaining a proper alignment of manufacturing structure with business strategy". However, this is not always easy to achieve for many companies since they will always have a legacy of physical investments that will be difficult to reverse even if the business strategy would prefer so.

Hayes and Pisano [5] argue that structural decisions tend to be centralized, in contrast to infrastructural decisions. However, many smaller structural decisions might need to be made throughout an organization in a decentralized way. Such decisions might accumulate and become as important as centrally made decisions in the long-term.

10.3 The Structural Prerequisites for How to Deal with Knowledge

How a company positions itself in the supply chain is a prerequisite for knowledge since to some extent it defines what the company sees as its core activity. The core activity directs innovations which are both incremental and more radical, and consequently it also directs knowledge creation.

Most of the literature emphasizes that companies should analyze the strategic effects of outsourcing and keep their core functions as core competency.

According to Mahmoodzadeh et al. [9] it is increasingly difficult for a global company to recognize what is core, and the most significant challenge is to reduce the risks in strategic outsourcing and to identify technologies that can enable the outsourcing process.

A company's positioning and relationship to its customers and suppliers are also important for knowledge transfer. Systems and formal structures, business culture, and other company contextual elements related to the company as an organizational unit will normally influence the ability to transfer tacit knowledge. Knowledge transfer tends to be more explicit between companies than within a single corporate or parent organization.

Manufacturing paradigms have been shifting from a focus within the walls of the manufacturing plant towards supply chain integration. This does not necessarily mean acquisitions and strategic alliances but integration of information systems for knowledge transfer. Concurrent engineering, supply chain integration, co-design between different organizations and customer relationship management have become the focus for processes that integrate customers, suppliers, product designers and manufacturing engineers and where they can work together to solve problems and explore new opportunities. This integration of knowledge resources represents an important difference between mass and lean manufacturing. In mass manufacturing there has been a tendency to in-source rather than outsource along the supply chain. Agile and adaptive manufacturing on the other hand places a focus on supply networks and regard knowledge sharing as a crucial element to create a business model for mass customization.

Process decisions concerning answers to questions such as, how automated a process should be, whether technologies should be developed in-house, technological leadership, and technologic fit between the product and the process, all represent issues related to knowledge creation. Knowledge creation is concerned with issues such as what knowledge needs to be created (the kind of processes we need knowledge about), how much knowledge is needed (if we aim to be technology leaders we need to focus a lot on knowledge), and who is needed for managing the knowledge process (if are we developing technologies internally). The need for knowledge transfer depends to a large extent on how intense the knowledge creation activities are, but in particular where they are taking place. If the knowledge creation is outsourced and/or centralized there will be challenges related to transfer and adapting knowledge.

Craft manufacturing, such as in the building of leisure boats, is typically oriented towards manual processes, and knowledge is created and transferred through 'learning by doing' and is thus classified mainly as tacit knowledge. In mass manufacturing, processes are more specialized and automated, with a centralized knowledge creation process dependent on explicit knowledge, while in lean manufacturing knowledge is much more oriented towards incremental and decentralized creation. Adaptive manufacturing is dependent on highly flexible processes and technologies which require high levels of knowledge creation, but also sharing and transfer of knowledge within the supply chain network.

The role of the manufacturing facilities might differ, especially according to the centralization–decentralization dimension. The 'centralized hub' will normally imply a centralized knowledge creation process, where for example a Project Management Office (PMO) has responsibility for the important parts of knowledge creation. The 'decentralized' and 'loose federation models' have a different distribution of the roles within knowledge creation, where for example a PMO mainly supports knowledge creation processes throughout the various functions. Decisions regarding location and capacity are closely related to vertical integration and the roles of the facilities. A very important issue related to location and outsourcing is knowledge, for example whether facilities are located within an innovative environment, such as described by Porter [10]. Also the physical distance between the facilities, especially the mother plant would also, despite technological enablers, have an impact on the knowledge creation process.

It is evident that the distribution of different roles among knowledge workers has an impact on the transference of knowledge. A centralized knowledge creation process would normally have disadvantages when it comes to capturing the tacit knowledge and making the centrally created knowledge, which often has to be formalized and explicit, and relevant for the different units. In a decentralized knowledge creation process the challenge is more likely to make the knowledge less contextual i.e., formalize the tacit knowledge or make it relevant in different contexts (still tacit knowledge) in other units.

Craft manufacturing is normally limited when it comes to scale and capacity, and hence the location of different facilities and transfer issues are not emphasized. In mass manufacturing there are more often multiple facilities, but normally within a centralized structure in knowledge creation both within the company and between the units. This centralized structure represents a transfer challenge from the mother plant to facilities, but also within the facilities. Lean manufacturing has moved towards a more decentralized structure, where knowledge creation to a large extent is addressed to the facilities and their relations to customers and suppliers. Knowledge is often tacit and related to incremental innovations. A challenge for lean manufacturing would then be to externalize the knowledge and make it relevant to the other units. Location decisions are often related to Just In Time (JIT) but also the ability to work concurrently on improvement and development with customers and suppliers. Adaptive manufacturing has a much more dynamic structure. Adaptive strategies are based on networks and a common knowledge base among many companies. This means complex processes related to knowledge creation and transfer between and within the units and networks involved.

10.4   Infrastructure, Exploiting the Structure

Hayes, Wheelwright and Clark [7] conclude that the key differences between winners and losers in the market place are that winners focus on creating value for customers, continuous improvement, quick adaptability to change, and extracting the full potential of human resources; they are what they term "learning organizations". This implies a focus on infrastructure which develops over time through persistent day-to-day practices, top management commitment, and cross-functional efforts to create capabilities and leverage the company structure [4]. In this respect, knowledge is a major part of the infrastructure. Coordination is necessary to secure the best effect following structure and infrastructure decisions, but in reality how these parts fit together is rarely encompassed by any strategic overview [4]. Knowledge cannot be managed in the same way as other kinds of resources [11] and although there exist a body of research on MNC (Multinational corporation) control and coordination (for reviews, see Martinez and Jarillo [12] and Doz and Prahalad [13]), there is a lack of research on the strategies to ensure knowledge transfer across different units [14, 15].

Skinner [4] asserts the importance of human factors and how a company develops the policies and systems that govern such activities such as capital budgeting, human resources, quality/process control, material flows, and performance measurement. He states that these infrastructural issues are often more important than structural decisions. He argues that such systems should be designed to encourage the continual adaptation and improvement of an organization's skill base rather than to achieve some 'optimal' strategic fit. Infrastructure or soft decisions relate to people, organizations and systems, and support the action element of a strategy. They typically involve middle managers on a more frequent basis than structural decisions. Walker [16], Bartlett and Ghosahl [17] and others emphasize the coordination challenge between, product, functional, and geographically oriented management when operating globally:

> The difficulty is further increased because the resolution of tension among the three management groups must be accomplished in an organization whose operating units are often divided by distance and time and whose key members are separated by barriers of culture and language [17, p. 513].

10.5 Structure and Infrastructure, a Complex Interplay of Decisions

It is often difficult to make a clear distinction between structural and infrastructural decisions. There is also a complex interplay between the different structural and infrastructural decisions, where in some instances the companies' culture, systems, and so on would require certain structural decisions about location of units, process choice, etc. This interplay between manufacturing structure and infrastructure is illustrated in Fig. 10.1. The concept of consistency is influenced by the ecology school of thought and suggests that the critical elements to be aligned to strategy are as follows [18]:

- *Internal to the firm*—focusing on strategy and organizational fit
- *External to the company*—a process seeking fit with the business environment
- *Internal–external fit*—internal formulation and implementation (i.e., developing infrastructure) are considered to be interactive elements.

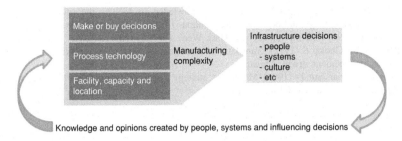

Fig. 10.1 The interplay of structure and infrastructure in manufacturing

Hill [3] stresses that a manufacturing structure and its complexity must be determined before infrastructure are developed. This is of course true when the heritage of facilities and other physical elements are considered.

References

1. Skinner W (1969) Manufacturing—missing link in corporate strategy. Harward Business Review, pp 136–145 (May-June)
2. Hayes RH, Wheelwright SC (1984) Restouring our competitive edge: competing through manufacturing. Wiley, NY
3. Hill T (2000) Operations management: strategic context and managerial analysis. Macmillan, London
4. Beckman S, Rosenfield D (2007) Operations strategy: competing in the 21st century. Irwin/McGraw-Hill, NY
5. Hayes RH, Pisano GP (1996) Manufacturing strategy: at the intersection of two paradigm shifts. Prod Oper Manage 5(1):25–41
6. Dangayach GS, Deshmukh SG (2001) Manufacturing strategy: literature review and some issues. Int J Oper Prod Manage 21(7):884–932
7. Chandler G (1996) Capacity planning and simulation. In: Walker J (ed) Handbook of manufacturing engineering (manufacturing engineering and materials processing). CRC, NY
8. Choe K, Booth D, Hu M (1997) Production competence and its impact on business performance. J Manuf Syst 16(6):409–4219
9. Mahmoodzadeh E, Jalalinia S, Yazdi FN (2009) A business process framework based on business process management and knowledge management. Bus Process Manage J 15(6):845–864
10. Porter ME (1980) Competitive strategy: techniques for analyzing industries and competitors. Free Press, NY
11. Von Krogh G, Ichijo K, Nonaka I (2000) Enabling knowledge creation. How to unlock the mystery of tacit knowledge and release the power of innovation. Oxford University Press, Oxford
12. Martinez JI, Jarillo JC (1989) The evolution of research on coordination mechanisms in multinational corporations. J Int Bus Stud 20(3):489–514
13. Doz YL, Prahalad CK (1993) Managing DMNCs: a search for a new paradigm I. In: Ghoshal S, Westney E (eds) Organizational theory and the multinational corporation. St Martin's Press, NY, pp 24–50
14. Bjorkman I, Barner-Rasmussen W, Li L (2004) Managing knowledge transfer in MNCS: the impact of headquarters control mechanisms. J Int Bus Stud 35:443–455

15. Foss NJ, Pedersen T (2002) Transferring knowledge in MNCs: the role of sources of subsidiary knowledge and organizational context. J Int Manage 8(1):1–19
16. Walker J (1996) Redemption revival. Mark Week 19(7):67–73
17. Bartlett CA, Ghosal S (1992) Transnational management, text, cases, and readings in cross-border management. Irwin, Chicago
18. Nath D, Sudharshan D (1994) Measuring strategy coherence through patterns of strategic choice. Strateg Manage J 14(1):43–61

Chapter 11
Make or Buy?

Abstract Outsourcing is essentially about the companies' position in the supply chain. There are many strategic reasons for outsourcing that despite the drawbacks have made it a trend in manufacturing. Outsourcing is normally viewed as involving the contracting out of a business function or process previously performed in-house to an external provider. This means that a company based on its strategic considerations will choose to split up some functions or processes and outsource them to an external organization mainly because they are not considered core business, or are perhaps too costly to perform in-house. Various support processes and functions are frequently not considered value adding for a manufacturing company. To maximize the advantages and minimize the risks of outsourcing the company needs to anchor these processes and decisions strategically. This means that the less tangible aspects such as control and knowledge need to be evaluated and dealt with in the strategic decision process. Of recent concern is the ability of businesses to outsource to suppliers outside of their national boundaries, sometimes referred to as offshoring or offshore outsourcing.

11.1 What is Outsourcing?

Outsourcing is about the companies' position in the supply chain. The literature refers to a situation for a company which is traditionally known as the 'make or buy' decision [1]. This is one of the most important strategic decisions of a manufacturing company as it affects most of its other strategic decisions. Outsourcing decisions can be complex, with difficult discussions and sometimes internal conflicts. The discussions have in some instances also involved governments and other public bodies at local or national level. The reason for this has been the consequences of buy versus make decisions on employment, and the roles of units. However, there are many compelling reasons for outsourcing that despite risks have made it to a trend in manufacturing.

A. Rolstadås et al., *Manufacturing Outsourcing*,
DOI: 10.1007/978-1-4471-2954-7_11, © Springer-Verlag London 2012

Fig. 11.1 In-house functions
and processes

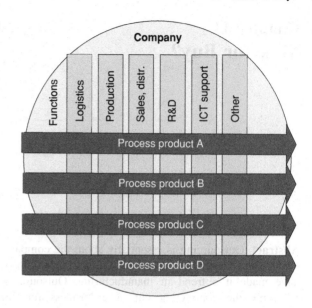

Outsourcing is normally viewed as involving the contracting out of a business function or process previously performed in-house to an external provider. This means that a company based on strategic considerations will choose to split up some functions or processes and outsource them to an external organization. These processes are not considered core business, or perhaps are too costly to have in-house. These are typically support processes and functions which are not considered to add great value to core transformation activities. We are also seeing outsourcing of some core transformation processes for reasons are also often related to flexibility, cost reductions, and so on. Outsourcing not only means purchasing products or services from sources that are external to the organization but also transfers the responsibility of physical business functions and often the associated knowledge to the external organization [2]. We also see examples of strategic outsourcing of R&D and other key activities under the argument of getting access to knowledge and innovations faster and cheaper.

Figures 11.1 and 11.2 illustrate how a company can outsource functions for example ICT support. We also see how the whole process, production, distribution and sales can be outsourced (Product C), and how the sub process of for example sales of Product D can be outsourced while the rest of the processes are kept in-house. The ownership of the outsourced functions or processes may differ. The company often keeps legal control of the units responsible for the outsourced functions and processes through ownership or joint ownerships. However, IT support is a typical example where the functions are outsourced without any kind of controlling ownership.

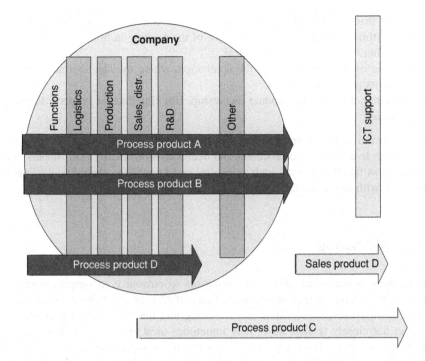

Fig. 11.2 Outsourced functions and processes

11.2 Why Outsource

There have been several studies that have examined the motivations for and benefits of outsourcing. Abraham and Taylor [3] identified the following: savings on wage and benefit payments; transfer of demand uncertainty to the outside contractor, and; access to specialized skills and inputs that the organization cannot itself possess. Kakabadse and Kakabadse [4] discusses similar motivations for outsourcing but identifies a greater emphasis on innovation and improvements; economies of scale; longevity of demand for the activity; quality; access to knowledge; skills; greater geographical coverage; and innovation where for example improvements can be made in quality through innovation, and where the development of new service products can lead to new demands.

In the literature the economic advantages of outsourcing are often taken for granted [5]. However, there can be several drawbacks and risks in outsourcing and companies should be aware of what can go wrong and what steps could be considered to mitigate these risks. Potential risks include:

- Communication costs especially in a cross cultural context
- Inadequate governance and management mechanisms
- Loss of control over crucial knowledge and technical staff

- Loss of leadership in business relations with subcontractors and stakeholders
- Underestimating backlash and resistance of the existing in-house team, but also from other stakeholders, i.e. political
- Dynamic of costs (salaries) of the contractors where the sub-contractor might change its pricing strategy
- Suppliers not respecting product ownership, IPR (Intellectual Property Rights), and so on.

To maximize the advantages and minimize the risks of outsourcing the company needs to anchor these processes and decisions strategically. This means that also the less tangible aspects such as control and knowledge should be evaluated and dealt with in the strategic way.

11.3 Off-Shoring

Of recent concern is the ability of businesses to outsource to suppliers outside national boundaries, sometimes referred to as off-shoring or offshore outsourcing. There is no universal definition of off-shoring and the term is in use in several distinct but closely related ways. It is sometimes used broadly to include substitution of a service from any foreign source for a service formerly produced internally to the firm. In other cases, only imported services from subsidiaries or other closely related suppliers are included. Kirkegaard [6] describes offshore outsourcing as "production relocation" i.e. instances of total or partial closure of an enter-prise's existing production units in one country with accompanying work force reductions. He also describes off-shore outsourcing as either the opening of affiliates abroad for the production of the same goods or services or the forging of a subcontracting contract with a nonaffiliated firm for production of the same goods or services in their home country. This means that off-shoring covers both the companies own units and units not owned by the company.

According to Sako [7, p. 3] off-shoring happens when:

> Private firms or governments decide to import intermediate goods or services from overseas that they had previously obtained domestically. It is therefore about sourcing decisions which involve (a) imports, (b) displacement of domestic production and associated jobs, and sometimes (c) foreign direct investment (FDI) outflows if sourcing happens from overseas affiliates.

The advantages and arguments for off-shoring are often similar of those for outsourcing, but companies are normally looking to take advantages of global differences in factor costs and competitiveness. These differences are illustrated in Chap. 5.

11.4 The Decision Process

Outsourcing decisions are structural decisions related to; positioning in supply chain; managing facilities and capacities; and also managing logistics through location. These strategic decisions also have major implications on infrastructure such as: human and organizational aspects; business culture; and knowledge and innovation processes. The pros and cons presented previously need to be considered in a holistic and transparent way.

Theories of decision and the decision making process is a wide and complex research area. An important dimension in decision making is to which degree decisions are deterministic. The Markov Decision Process (MDP) is a model of system dynamics in sequential decision problems that involves probabilistic uncertainty about future states of the system [8]. It contains:

- a set of possible world states (S)
- a set of possible actions (A)
- a real valued reward function (R)
- a description of each action's effects in each state (T).

An important assumption in MDP is the assumption that the effects of an action taken in a state depend only on that state and not on the prior history. The decision maker (agent) observes the state and performs an action accordingly. The system then makes a transition to the next state and the agent receives some reward.

There are many cognitive biases that question the relevance of deterministic models like MDP, especially in strategic decision processes such as outsourcing decisions. Das and Teng [9] describe four types of cognitive biases:

- Focusing on limited targets
- Exposure to limited alternatives
- Insensitivity to outcome probabilities
- Illusion of manageability

These biases will provide different modes of decision making including: rational, avoidance, logical incrementalist, political and garbage can [9].

Non-deterministic policies are inherently different from stochastic policies since stochastic policies assume a randomized action selection strategy with some specific probabilities, whereas non-deterministic policies do not impose such constraint. We can thus use best-case and worst-case analysis with non-deterministic policies to highlight different scenarios with the human user. Several models have been developed for how companies could deal with these kinds of biases or have a structured non-deterministic process. One such aspect is how to deal with the situation where companies in practical cases do not have complete knowledge of the system at hand. Fard and Pineau [10] present a way of applying MDP in non-deterministic decision process. They describe how companies instead, may get a set of trajectories collected from the system according to some specific policy. In many cases, the data may be too sparse and incomplete to uniquely identify the best option and near-optimal solutions.

It is also important to have in mind that strategic decisions processes like outsourcing and location decisions normally are or would benefit from being group decisions. Differences in individual preferences and the dynamics within group and group processes represent a complicating element of both deterministic and non-deterministic models.

In the outsourcing decision process, the transaction cost approach has frequently been used. This approach is based on theory of firms and markets that, in its basic form, gives context to how business is conducted. Their transactions are organized either internally or through markets, depending on the costs to organize those activities [1].

Besides the acquisition cost, specific asset investments that not only are more efficiently organized within the company, but provide a source of long-term returns and hence strategic value are examined deeper. If a particular production process produces value and strategic advantage the company would probably benefit in the long turn to keeping this process in-house.

References

1. Coase RH (1937) The nature of the firm. Economica 4(16):386–405 (New Series, Nov 1937)
2. McCarthy I, Anagnostou A (2004) The impact of outsourcing on the transaction costs and boundaries of manufacturing. Int J Prod Econ 88:61–71
3. Abraham KG, Taylor SK (1993) Firms' use of outside contractors: theory and evidence, NBER working paper, 4468, Cambridge, MA
4. Kakabadse N, Kakabadse A (2000) Critical review—outsourcing: a paradigm shift. J Manage Dev 19(8):670–728
5. Berggren C, Bengtsson L (2004) Rethinking outsourcing in manufacturing: a tale of two telecom firms. Eur Manage J 22(2):211–223
6. Kirkegaard JF (2007) Offshoring, outsourcing and production relocations—labor market effects in the OECD and developing Asia, working paper Peterson Institute for International Economics, Washington D.C.
7. Sako M (2005) Outsourcing and offshoring: key trends and issues. Paper presented at the emerging markets forum. Oxford, UK
8. Bellman R (1957) Dynamic programming. Princeton University Press, NJ
9. Das TK, Teng B-S (1999) Cognitive biases and strategic decision processes: an integrative perspective. J Manage Stud 36(6):757–778
10. Fard MM, Pineau J (2011) Non-deterministic policies in Markovian decision processes. J Artif Intell Res 40:1–24

Chapter 12
The Geographical Footprint

Abstract Where to locate manufacturing units have always been among the most important decisions in a manufacturing strategy and is to a large extent independent of whether the company has outsourced some operations or are doing them inhouse. Depending on level of decentralization of processes and responsibilities, the location is a physical manifestation and a footprint of the manufacturing strategy. These decisions require a minimum of facts to allow quantified cost benefit analysis. However, the increasingly important intangible aspects, for example related to innovation and improvement must be captured in the decision making process. There are different methods to assess, analyze and compare alternatives. An example is the Analytical Hierarchy Process (AHP) but there are also many other techniques for organizing and analyzing complex decisions. The geographical footprint is about what kind of activities the different units have in the production system, the role of the units and the relation between them. The roles of the units should not only be a prerequisite for decisions on whether to outsource or not. It also defines challenges and requirements to coordination activities for example related to innovation, improvements and knowledge transfer.

12.1 Location of Facilities

Manufacturing is becoming more distributed and globalized. This has until recently been explained exclusively by the need for proximity to resources and markets. The emerging knowledge-based economy with its profound effects on markets, society, and technology has accelerated this process and shifted the focus away from proximity towards access to knowledge resources, in particular knowledge workers.

Where to locate units have always been among the most important decisions in a manufacturing strategy and is to a large extent independent of whether the company has outsourced particular processes or are carrying them out inhouse.

A. Rolstadås et al., *Manufacturing Outsourcing*,   101
DOI: 10.1007/978-1-4471-2954-7_12, © Springer-Verlag London 2012

Depending on level of decentralization of processes and responsibilities, the location is a physical manifestation and a footprint of the manufacturing strategy. This manifestation represents important, long-term costs and commitments that are not easily changed or removed.

The current literature on structural issues relating to facilities and outsourcing can be classified into two groups according to their focus [1]: the first group is choice of organizational forms, focusing on choices between outsourcing and integration, and choice between home and off-shore localization [2–5]. The second group of research is concerned exclusively with, evaluation and selection of location [6–8].

12.2 Approaches to Location and Outsourcing Decisions

Location issues have been approached with varying assumptions and perspectives like economic geography, international business, management science, and others. Geographers have tempted to present holistic approaches to location decisions through agglomeration and cluster theories. Porter [9, p. 254] defines clusters as "geographically proximate groups of interconnected companies and associated institutions in a particular field, linked by commonalities and complementarities". One of the common features of these theories is their implication that regional productivity should depend positively on the regional density of economic activity [10]. The reasons for this are that new technologies are adopted and exchanged more rapidly in places of dense economic activity. Another reason for agglomeration effects is related to the size of markets and specialization economies.

Location theory distinguishes three different sets of factors driving the firm's location decision problem: external economies, costs of production factors, and accessibility (transportation costs) to demand markets [11]. Ellison and Glaeser [12] are among the authors that have developed location indexes focusing on these factors. The company-level and microeconomic factors were not emphasized in the research of international manufacturing until the 1960s [13]. Since then we have seen the emergence of internationalization schools [14]:

- *Uppsala Model/U-model*—companies base their internationalization on experimental knowledge and expand their activities to markets and countries with "close psychic distance"
- *Innovation-Related Model/I-model*—internationalization as a process with gradual stages starting with sporadic exports
- *Eclectic Model*—the widely accepted model for the study of location decisions in the international business literature. The focus is on market imperfections and transaction cost in location decisions
- *International Marketing and Purchasing Group/IMC-model*—focusing on relationship issues, especially to customers and suppliers
- *Business Strategy*—internationalization as a strategic interplay of market opportunities, resources, and managerial style.

All these schools of thought represent theories and strategic aspects of relevance for manufacturing. However, it is an open question to which extent companies utilize a systematic approach able to capture the strategic manufacturing aspects when deciding the location of these units. We believe that in paradigms of today such as in lean manufacturing, location decisions must be based on key aspects of these paradigms for example, employee involvement, continuous improvement, and supply chain collaboration. For a truly lean company, the best location could be in an environment where employees are used to working proactively with quality improvement, where key customers are located, where customer feedback is accountable, where first tier suppliers and R&D partners are located, and so on. A business culture, where local managers are able to foresee trends and work internally to change direction or roles of a manufacturing unit, could be a crucial location aspect of any location decision [14].

The outsourcing, location and partner selection decisions are crucial and require a good and transparent process well anchored in the overall strategy process. These decisions require a minimum of facts to allow quantified cost benefit analysis. However, the increasingly important intangible aspects for example related to innovations and improvement must be captured in the decision process. There are different methods to assess and analyze and compare alternatives. An example is Analytical Hierarchy Process (AHP), but there are also many other criteria based, structured technique for organizing and analyzing complex decisions.

12.3 Focus on Process or Product

The geographical footprint of manufacturing is not only a question of location and partner selection. The main issues are also related to what kind of activities the different units have in the production system, the role of the units and the relationships between them. The roles of the units should not only be a prerequisite for deciding whether to outsource or not. It also defines challenges and requirements for coordination activities, for example related to innovation, improvements and knowledge transfer.

The structural decisions regarding the manufacturing facilities of a global firm are interdependent. However, a review of the relevant literature reveals that facility location, sizing, and technology selection decisions can be dealt with separately. Dou and Sarkis [1] exemplify this through the lack of integrated models for location factors and supplier selection factors.

The decision whether to focus plants by product lines or by processes is an important strategic decision. The evaluation might be as follows: if different products are sold in markets that require different competitive priorities (e.g., low cost in one, innovation in another), then focusing along product lines would then be preferable [15]. Different facilities could then tailor their equipment, operating policies, and knowledge to fit their particular markets. In contrast, if different product families compete in roughly similar ways (e.g., low cost), but the processes are very different between segments, the operating policies and knowledge required would also differ, in which case a process focus would be preferable. How to focus and organize the different units and production system is an important decision not only for location of functions and responsibilities, but also to set direction for innovation and improvement processes, hence the knowledge creation and transfer.

12.4 Where to Do What

Most large companies have some kind of heritage or history going back to one specific unit and location from where the company started its growth to becoming a multi-site and global company through organic growth and/or acquisitions. The original location, the 'mother site', often remains as one of the most important units in the company, for example with headquarters and R&D functions. However, there are many other ways of organizing the units, their roles, functions and relations. What was the original mother site might not keep such a central role over time and other units might become the 'mothers' of a production system consisting of more than one mother-satellite structures. The roles of the different units in a company or production system vary according to many dimensions, but centralized versus decentralized decision making is of particular importance.

Centralized versus decentralized decision making could be illustrated through the different organizational configuration of "mother plant—satellite manufacturing" The term is not very common in the manufacturing literature, but international management literature may prove helpful in this respect. Bartlett and Ghoshal [15] make the following distinctions, focusing on companies' "mentality":

- *International*—companies regard themselves as fundamentally domestic, with some foreign appendages. Decisions related to foreign operations tend to be made in an opportunistic and ad hoc manner
- *Multinational*—companies recognize and emphasize the differences among national markets and operating environments. Their strategies are founded on the multiple, nationally responsive strategies of the companies' worldwide subsidiaries
- *Global*—companies think in terms of creating products for a world market and manufacturing them in a few highly efficient plants, often at the corporate center. In such companies, R&D is typically managed from the headquarters, and most strategic decisions are also taken there
- *Transnational*—companies are more responsive to local needs while retaining their global efficiency. Resources and activities are dispersed but specialized, so as to achieve efficiency and flexibility at the same time. These dispersed resources are integrated into an interdependent network of worldwide operations.

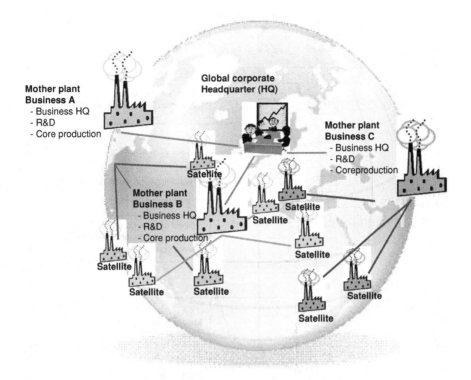

Fig. 12.1 Mother plant satellite manufacturing of the multi-home based company

A 'mother plant—satellite structure' resembles the structure found in both global and transnational companies. Even in a decentralized structure, such as that of a transnational company, there will normally be an administrative heritage to one location. Sölvell [16] has presented an application involving a transnational model with several 'mother plant—satellite structures', and a coordinating corporate unit. Such a model for a multi- home based manufacturing structure is illustrated in Fig. 12.1.

As we can see, the coordination challenges become important in a production system that includes many satellites. However, organizing the activities through business units and mother plants could reduce some of coordination challenges for the corporate management.

Figure 12.2 shows Bartlett and Ghosahl's [17] three basic organizational configurations that are all relevant in a 'mother plant—satellite manufacturing' structure. The configurations are focusing on flows, processes, and control dimensions.

Even if there is a basic structure, as illustrated in Fig. 12.1, the different facilities might have different roles. Ferdows [18] defines six roles which manufacturing facilities could have in a distributed structure that define their strategic role:

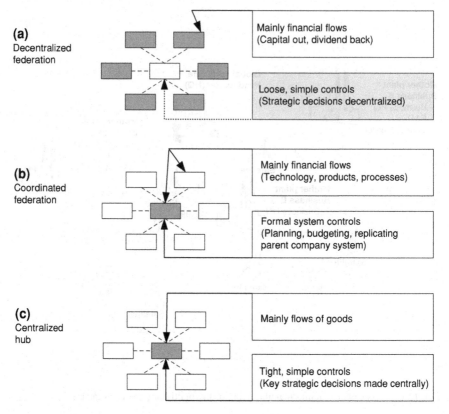

Fig. 12.2 Organizational configuration model [17]

- *Offshore factory*—established to manufacture specific items at low cost, which are then exported for further work or sale. Investments in technical and managerial resources are kept at a minimum, and price and supplier selection are decided centrally
- *Source factory*—tends to be located where manufacturing costs are relatively low, infrastructure is relatively developed, and the workforce is skilled. The local managers make more decisions on planning and operational matters, but also related to tactical decisions on process improvement and process redesign
- *Server factory*—supplies defined geographical markets and is typically a means to overcome national entry/trade barriers and logistics challenges. It has more local authority than an offshore factory, but decisions are mainly related to local modifications
- *Contributor factory*—serves defined geographical markets, but the responsibilities extend to process and product development, with close relations also to suppliers (selection and development)

- *Outpost factory*—has as its primary task to collect information. Such factories are normally located in clusters where there are advanced suppliers, customers or competitors
- *Lead factory*—creates products, processes, and technologies to be used by the whole company. The factory will normally be authorized to 'tap' other units for knowledge and resources (personnel) and make decisions for the whole corporation.

These organizational configurations do not necessarily apply only for units owned by the mother plant. Outsourced units might also have strong or loose coordination mechanisms or roles and strategic positions as described above. A parent companies' ability to define the role of the outsourced functions and processes, or the role of an external partner, are much more limited than if these functions or processes were done in-house. Even if there are formal regulations that can be applied between the partners a good and constructive partnership requires other mechanisms that could create collaboration through the distribution of roles and responsibilities between the units and partners.

Learning and the transfer of knowledge between units have become increasingly important in building and maintaining knowledge. The companies not only have to promote the flow of knowledge but also maximize knowledge creation and innovation [19]. Where decisions on improvement and development are located will be important for the role within the 'mother plant—satellite structure'. These tactical processes are increasingly important for companies aiming to adapt to and explore opportunities in the business environment [20].

References

1. Dou Y, Sarkis J (2008) A joint location and outsourcing sustainability analysis for a strategic offshoring decision. Available at SSRN: http://ssrn.com/abstract=1125496
2. McLaren JE (2000) Globalization and vertical integration. Am Econ Rev 90(5):1239–1254
3. Grossman GM, Helpman E (2005) Outsourcing in a global economy. Rev Econ Stud 72(1): 135–159.
4. Antràs, Helpman (2004). The second group of research has been concerned with location factors, evaluation and selection [6–8]
5. Ruiz-Torres AJ, Mahmoodi F (2008) Outsourcing decisions in manufacturing supply chains considering production failure and operating costs. Int J Supply Chain Manage 4(2):141–158
6. De Almeida AT (2007) Multicriteria decision model for outsourcing contracts on utility function and ELECTRE method. Comput Oper Res 34(12):3569–3574
7. Chan FTS, Kumar N (2007) Global supplier development considering risk factors using fuzzy extended AHP-based approach. Omega 35(4):417–431
8. Porter ME (2000) Locations, clusters, and company strategy. In: Clark GL, Feldman MP, Gertler MS (eds) The Oxford handbook of economic geography. Oxford University Press, New York
9. Ciccone A, Hall RE (1996) Productivity and the density of economic activity. Am Econ Rev 86(1):54–70

10. Guimaraes P, Figueiredo O, Woodward D (2004) Industrial location modeling: extending the random utility framework. J Reg Sci 44(1):1–20
11. Ellison G, Glaeser E (1997) Geographic concentration in U.S. manufacturing industries: a dartboard approach. J Political Econ 105(5):889–927
12. Hymer S (1960) The international operations of national firms: a study of direct investment. PhD thesis, MIT, Published by MIT Press under the same title in 1976
13. Salgado O (2004) Assessing the process of localization of parts in a Mexican car manufacturer and its effect on regions growth. Institute for Manufacturing, University of Cambridge. http://www.ifm.eng.cam.ac.uk/people/os242/documents/localization.pdf
14. Hayes RH, Pisano GP (1996) Manufacturing strategy: at the intersection of two paradigm shifts. Prod Oper Manage 5(1):25–41
15. Bartlett CA, Ghosal S (1992) Transnational management, text, cases, and readings in cross-border management. Irwin, Chicago
16. Sölvell Ö (2002) The multi-home based multinational–combining global competitiveness and local innovativeness. Paper presented at the symposium in honor of John Stopford London, 16–18 June 2002
17. Bartlett CA, Ghosal S (1998) Transnational management, text, cases, and readings in cross-border management. Harvard Business School Press, Boston
18. Ferdows K (1997) Making the most of foreign factories. Harward Bus Rev Mar–Apr: 73–88
19. Nohria N, Ghosal S (1997) The differentiated network: organizing multinational corporations for value creation. Jossey-Bass Publishers, San Francisco
20. Henriksen B, Andersen B (2010) Is there a tactical level of business processes?: emphasizing processes that enable adaptability, change, and improvement. TQM J 22(5):516–528

Chapter 13
Approaching the Partner Selection and Location Decision

Abstract Criteria-based methods can be used to capture key elements for strategic decisions. The criteria and their weight can then differ according to the paradigms in which they are applied. This is also reflected in differences in the knowledge creation processes. Craft manufacturing for example is normally limited when it comes to scale and capacity, and hence the location of different facilities and knowledge transfer issues need not be emphasized. In mass manufacturing, there are frequently multiple location facilities, but normally within a centralized structure for knowledge creation both within the company and between the dispersed units. This centralized structure represents a transfer challenge from the mother plant to facilities, but also within the facilities. Lean manufacturing has moved towards a more decentralized structure, where knowledge creation to a large extent is addressed by the various facilities and their relationships with customers and suppliers. However, companies having a manufacturing strategy based on principles from sustainable manufacturing have a complex set of decisions to make when it comes to outsourcing, location of units, and so on. Even if the company has a strategy for vertical integration, it is normally difficult to own the product lifecycle.

13.1 Manufacturing Paradigm as Premise for Outsourcing Decisions

The mother plant—satellite structure implies a minimum level of activities and processes concentrated at one facility, the mother plant. However, as shown in Fig. 12.2, the role of the facilities might differ, especially according to the centralization–decentralization dimension. The 'centralized hub' will normally imply a centralized knowledge creation process, where for example a project management office (PMO) has responsibility for the important parts of knowledge creation.

A. Rolstadås et al., *Manufacturing Outsourcing*,
DOI: 10.1007/978-1-4471-2954-7_13, © Springer-Verlag London 2012

A project management office (PMO) could be defined as a competence center created to integrate project management and project-based activities in an organization. If, where, and how a company establishes a project management office are structural questions with major knowledge implications.

The decentralized and 'loose federation models' have a different distribution of the roles within knowledge creation where for example a PMO mainly supports knowledge creation processes throughout the units.

Decisions regarding location and capacity are closely related to vertical integration and the roles of the facilities, but another very important issue related to location is knowledge creation, for example whether facilities are located within an innovative environment, such as described by Porter [1]. Also the physical distance between the facilities, especially the mother plant–satellite would also, despite technological enablers, have an impact on the knowledge creation process.

It is evident that the distribution of roles has an impact on the transfer of knowledge. A centralized knowledge creation process would normally have disadvantages when it comes to capturing the tacit knowledge and making the centrally created knowledge, which often has to be formalized and explicit, relevant for the different units. In a decentralized knowledge creation process the challenge is more likely to make the knowledge less contextual—i.e., formalize the tacit knowledge—or make it relevant in different contexts (still tacit knowledge) in other units.

Craft manufacturing is normally limited when it comes to scale and capacity, and hence the location of different facilities and transfer issues are not emphasized. In mass manufacturing there are more often multiple facilities, but normally within a centralized structure in knowledge creation both within the company and between the units. This centralized structure represents a transfer challenge from the mother plant to facilities, but also within the facilities. Lean manufacturing has moved towards a more decentralized structure, where knowledge creation to a large extent is addressed to the facilities and their relationships to customers and suppliers. Knowledge is often tacit and related to incremental innovations. A challenge for lean manufacturing would then be to externalize the knowledge and make it relevant to the other units. Location decisions are often about JIT (Just In Time) but also the ability to work concurrently on improvement and development with customers and suppliers. Adaptive manufacturing has a much more dynamic structure. Adaptive strategies are based on networks and a common knowledge base among many companies. This means that there are complex processes related to knowledge creation and transfer between and within the units and networks.

13.2 Location Decisions and Partner Selection

A company with facilities which are geographically dispersed will have a number of advantages, one of which will be its flexibility to operate as a network. The activities have to be coordinated in order to be competitive, and the centralized management of activities is often referred to as the way to accomplish this. Ensign [2] points to a strong central leadership role, as particularly important, especially for companies operating globally. This implies a strong role of the 'mother' in a mother plant—satellite structure. Frost [3] is opposed to this view and finds empirical evidence to support the claim that a geographically dispersed global company functions not only to exploit, but increasingly to serve.

The coordination challenges are to a large extent decided when decisions on location are made. Thus, roles, process choice, capacities, etc. are important prerequisites for coordination and underline the importance of making the strategic outsourcing and location decisions in a systematic and transparent way.

Global location issues have been approached in different disciplines such as economic geography [4], management science and international business [5]. Hayter [6] identifies three approaches to facility decisions:

- *Neoclassical*—choose facilities to locate as a result of cost minimization and profit maximization. Facility location problems are one of the most common applications of optimization methods. For an overview of research, models and literature (see, for example, Plastria [7])
- *Behavioral*—considers the assumptions about rationality and perfect information unrealistic. The location is interpreted as a decision-making process, based on a company's perception and evaluation for an "information bed" [6]
- *Institutional*—critique of the neoclassical, and also the behavioral approaches, due to their implicit assumption of a static environment [8]. This approach emphasizes the strategy and aspects shown in Chap. 3.

Systematic approaches aiming to evaluate location decisions exist. The index proposed in Ellison and Glaeser [9] is established as a preferred method for measuring location of economic activity focusing on the systematic forces that lead to spatial concentration (for example, wages, land costs, market, accessibility, and transportation costs). Other authors, such as Guimarães et al. [10], have reviewed and further developed these methods.

The classical facility location problem has been defined as an optimization problem [11]. Cost/benefit analysis and different kinds of net value calculations seems as the most common methods providing the foundation for the concrete location decisions, occasionally accompanied by risk analysis (for example Belli et al. [12]). Cost/benefit analysis enables comparisons of locations and alternative use of the resources. The constraints of these analyzes are obvious as they just capture the accountable (and hardly even that) aspects, and with a very limited timeframe. Our concern is that the increasingly emphasized, but not easily accounted for, aspects of strategies such as innovation, knowledge, stakeholder relations, etc. are not covered in a systematic way. These kinds of aspects are often discussed and considered in the process of location decisions, but mainly based on diffuse criteria, subjectivity, and not through a systematic approach. We do not question the importance of quantified cost/benefit analysis that enable us to document expected return on investments, but we argue for a systematic way of capturing the less measurable aspects of location decisions. The Analytical Hierarchy Process (AHP) is a criteria-based method developed by Saaty [13] that could be a relevant framework and has been popular among location consultants [14]. However, AHP has mainly been applied using conventional tangible criteria such as tax, cost, market proximity, etc. [15].

A structured approach would be necessary to assure that a company makes location decisions that are in line with their overall strategies and basic manufacturing principles. The approaches and indexes from geographers and location theorists might give important input to decisions and decision criteria regarding location, but the more structured methods illustrated in Fig. 13.1 have the strongest links to project management literature [16]. Two complementary approaches to facility locations have emerged [17], one focusing on optimization of the number, size and geographic configurations, while more recent literature increasingly focuses on customer and supplier relations.

We also see combinations and extensions of methods. One example is an association of geographical information systems (GIS) and criteria-based decision methods [18]. The combination of approaches and methods could be enabled by stepwise approaches, with a primary sorting followed by more detailed analysis, for example cost/benefit analysis of a few alternatives.

Location methods can be grouped into "direct evaluation methods" and "criteria-based methods" [16]. In direct evaluation, alternatives are evaluated and compared, often based on experience, intuition or beliefs, while in the criteria-based methods the alternatives are evaluated according to several criteria.

The criteria-based methods allow decision-makers to model complex and strategic issues, such as location, based on objectives (criteria) for several alternatives. These methods are making evaluations according to a scale, and the scores are normally illustrated in quantitative terms. They represent a framework to cope with multiple criteria situations involving intuitive, rational, qualitative, and quantitative aspects [15]. The criteria will normally have different importance and a weighting scheme will be important. When the scores are multiplied with the weights of the criteria and added together, a weighted score for each alternative emerges. Figure 13.1 illustrates how these methods could capture important aspects of the manufacturing paradigms in outsourcing decisions.

If the less tangible aspects are covered in the location decision process then companies have a better foundation for capturing the coordination challenges they will meet. In this way companies would also be able to identify solutions or enablers for coordination. To which extent decisions and responsibilities are decentralized or centralized are important strategic issues not only because they might be vital according to the manufacturing paradigm they are attached to, but also because this could represent both challenges and solutions for coordination of units.

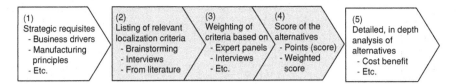

Fig. 13.1 Stepwise approach for a criteria-based method

13.3 Strategic Criteria in Location and Outsourcing Decisions: An Example

When outsourcing decisions and decisions around location, roles, etc. are to be made, the basic manufacturing principles reflected in the relevant manufacturing paradigm need to be captured in the decision process. Criteria-based methods such as the AHP or 'scoreboard' methods could be essential for assuring a robust decision process. Scoreboard is a method that is easily applicable using subjective criteria [16]. In these methods, almost every aspect that could have an importance for the location can be considered and filtered through stepwise approach as illustrated in Fig. 13.1.

It is important that the link to the overall strategic requisites is established. This requires a strategy process where for example basic manufacturing principles are defined and anchored in the overall corporate strategy (Step 1). There are several ways that the relevant location criteria can be identified. This could be done through an analytical process, where experts/key persons extract the criteria from the strategies (for example based on theory), through interviews, brainstorming sessions, etc. (Step 2). In step 3 it is important to define and describe the criteria in a coherent way that makes it possible to have a common understanding of them when they are weighted. There are several ways to weight the criteria, but interviewing key persons in the company will normally give us a foundation according to a common understanding of the strategic impact. In step 4, the different location alternatives are assigned scores for each criterion, and the weighted scores make it possible to compare the alternatives in a quantified manner (Step 5). A risk assessment of the scores could extend the model into a step 6.

Step 2–4 can be regarded as a strategic check on the alternatives, but the weighting process only captures what the persons conducting the weighting perceive as the strategy, and not necessarily what is defined as the strategy. In this way, a process as described in Fig. 13.1 could also give us a hint of how the managers, employees, and other stakeholders understand the companies'—a sort of reality check of the strategy.

AHP is a more complex method, where the alternatives are evaluated according to a hierarchy of objectives and goals, which represent the criteria at different levels. This hierarchy is often based on a top–down approach where objectives and goals are predefined from the top management. The criteria-based method illustrated in Fig. 13.1 is more likely to represent a bottom–up approach, especially

Table 13.1 Location criteria, example from the automotive industry

Group	Area	Criteria		Weight	Alternative 1		Alternative 2		Alternative n	
			Criteria	SUM	Score	Weighted	Score	Weighted	Score	Weighted
Structure	1.1 Supply chain	1.1.1	Market access	47						
		1.1.2	Transaction costs (local)	25						
		1.1.3	Cross border transaction costs (exclusive tax, etc.)	25						
		1.1.4	Risk (e.g. quality–defects)	45						
		1.1.5	Global sourcing capacity	24						
		1.1.6	Strategic partners	12						
		1.1.7	Flexibility(1)—easy to "switch" local suppliers	19						
		1.1.8	Flexibility(2)—easy to change functions/roles	24						
		1.1.9	Flexibility(3)—easy to adjust/change capacity	23						
			SUM	244						
	1.2 Facilities	1.2.1	Existing(own) unit, roads, ports, telecom, etc.	26						
		1.2.2	Local sourcing: raw materials, components, etc.	30						
		1.2.3	Local sourcing: equipment and operational services	27						
		1.2.4	Other	6						
			SUM	89						
	1.3 Process and process technology	1.3.1	Quality/cost manual processes	45						
		1.3.2	Quality/cost automated processes	39						
		1.3.3	Workforce stability	48						
		1.3.4	Access to new technologies	27						
		1.3.5	Educational level (relevant for products/processes)	33						
		1.3.6	Adaptability for new technologies	18						
		1.3.7	Governmental incentives-manual processes	13						
		1.3.8	Governmental incentives-automated processes	13						
		1.3.9	Other	6						
			SUM	242						

(continued)

Table 13.1 (continued)

Group	Area	Criteria for location decision		Weight	Alternative 1		Alternative 2		Alternative n	
			Criteria	SUM	Score	Weighted	Score	Weighted	Score	Weighted
Infrastructure	2.1 Sourcing	2.1.1	Strategic sourcing partners	17						
		2.1.2	Local collaborative environment	13						
		2.1.3	Opportunity for shared service (centers)	7						
			SUM	37						
	2.2 Management and development	2.2.1	Business culture similar to "mother plant"	13						
		2.2.2	Knowledge in mass manufacturing	22						
		2.2.3	Knowledge in lean manufacturing	35						
		2.2.4	Risk—political stability, etc.	33						
		2.2.5	Other	12						
			SUM	115						
	2.3 Business clusters and agglomerations	2.3.1	Strategic partners (e.g. presence expected by them)	42						
		2.3.2	"New customers"-new business opportunities	36						
		2.3.3	Strategic R&D partners	14						
		2.3.4	Strategic suppliers	19						
		2.3.5	Presence of competitors	29						
		2.3.6	Strong "cluster relations"	10						
			SUM	150						

through the weighting process (step 3) that could involve people from different parts and levels of the organization. This process is likely to motivate strategy discussions and a weighting of criteria that is more in line with the real strategy.

Table 13.1 is an example from the automotive industry where a company was in a process of choosing location for manufacturing units. The company was attached to basic lean principles which should be reflected in the location criteria and the weighting of them made by senior managers in the company. The method was applied for the first scanning of potential locations and based on the weighted score the company could set up a short list of locations.

13.4 Location Criteria and Sustainable Manufacturing

Lean manufacturing and its focus on resource efficient manufacturing, waste reduction, and so on has close links to sustainable manufacturing. Consequently criteria reflecting lean manufacturing are also relevant for companies aiming for sustainable manufacturing.

Companies having a manufacturing strategy based on principles from sustainable manufacturing have a complex set of decisions to make when it comes to outsourcing and location of units. Even if the company has a strategy for vertical integration, it is normally impossible to 'own' the entire product lifecycle.

Thus, sustainable manufacturing's focus on product lifecycle requires analytical skills and supporting methods to assure a maximum of knowledge about all parties involved in the product lifecycle. There are several methods for life cycle management and life cycle assessment. Criteria-based methods as shown in Fig. 13.1 and Table 13.1 could reflect stronger focus around premises such as transport (emission) and energy sourcing (i.e. renewable), disposal handling, and governmental incentives for environmental friendly actions.

References

1. Porter ME (1980) Competitive strategy: techniques for analyzing industries and competitors. Free Press, New York
2. Ensign PC (1999) The multinational corporation as a coordinated network: organizing and managing differently. Thunderbird Int Bus Rev 41(3):291–322
3. Frost AS (1998) The geographic sources of innovation in the multinational enterprise: US subsidiaries and host country spillovers, 1980–1990. PhD Thesis, MIT, Cambridge, MA
4. Porter ME (2000) Locations, clusters, and company strategy. In: Clark GL, Feldman MP, Gertler MS (eds) The Oxford handbook of economic geography. Oxford University Press, New York
5. Bartlett CA, Ghosal S (1998) Transnational management, text, cases, and readings in cross-border management. Harvard Business School Press, Boston
6. Hayter R (1997) The dynamics of industrial location. Wiley, New York

7. Plastria F (2001) Static competitive facility location: an overview of optimisation approaches. Eur J Oper Res 129(3):461–470

8. Brouwer AE, Mariotti I, Van Ommeren JN (2004) The firm relocation decision: an empirical investigation. Ann Reg Sci 38:335–347

9. Ellison G, Glaeser E (1997) Geographic concentration in U.S. manufacturing industries: a dartboard approach. J Political Econ 105(5):889–927

10. Guimaraes P, Figueiredo O, Woodward D (2004) Industrial location modeling: extending the random utility framework. J Reg Sci 44(1):1–20

11. Drezner Z, Hamacher HW (eds) (2002) Facility location: applications and theory. Springer, Berlin

12. Belli P, Anderson JR, Howard N, Barnum HN, Dixon JA, Tan JP (2001) Economic analysis of investment operations: analytical tools and practical applications. The World Bank, Washington, pp 73–82

13. Saaty TL (1980) The analytic hierarchy process, planning, priority setting, resource allocation. McGraw-Hill, New York

14. Viswanadham N, Kameshwaran S (2007) A decision framework for location selection in global supply chains. In: IEEE international conference on automation science and engineering, CASE 2007

15. Alberto P (2000) The logistics of industrial location decisions: an application of the analytic hierarchy process methodology. Int J Logistics Res Appl 3(3):273–289

16. Rolstadås A (2001) Praktisk prosjektstyring. Tapir Akademisk Forlag, Trondheim

17. Rich N (1995) Quality function deployment: a decision support matrix for location determination. In: 2nd International symposium on logistics, Nottingham, 11–12 July, pp 295–300

18. Agouti T, Tikniouine A, Eladnani M, Elfazziki A (2007) Toward an Integration of the fuzzy Logic and MCDA to GIS: application to the project of the localization of a site for the implantation of chemical products factory. In: Computational intelligence and intelligent informatics, 2007. International symposium on ISCIII apos;07, 28–30 Mar 2007, pp 79–83

Chapter 14
Dealing with Complexity: Infrastructure Decisions

Abstract The interplay of different strategic decisions such as outsourcing, and the complexity and changing business environment create a strategic context where coordination and collaboration become critical issues in manufacturing. These are infrastructural issues not only about long term strategy related to markets, products and processes but also about planning and development (tactical decisions) and continuous improvement in products and operations. The coordination challenges are functions of the complexity of processes, technology, and so on, but also of the role, number and size of the different units. The degree of autonomy and spatial and cultural distance are but some of the challenges of coordination. In this picture, the outsourcing decisions appear particularly important, and coordination mechanisms and other infrastructure issues need to be integrated as part of the outsourcing decision. Agency theory has been important in facilitating discussions around better coordination. Mintzberg associates coordination mechanisms to organizational factors such as specialization, decentralization, formalization, size, age (maturity), environment, and power. Coordination mechanisms can be grouped into four categories: structure and formal; informal; hybrid; internal quasi-markets and; internal prices.

14.1 Outsourcing Need to be Accompanied by Coordination Mechanisms

As we saw earlier in Fig. 3.2 the interplay of different strategic decisions such as outsourcing, and the complexity and changing business environment create a strategic context where coordination and collaboration become critical issues in manufacturing. These are infrastructural issues not only about long term strategy related to markets, products and processes but also about planning and development (tactical decisions) and continuous improvement in products and operations.

A. Rolstadås et al., *Manufacturing Outsourcing*,   119
DOI: 10.1007/978-1-4471-2954-7_14, © Springer-Verlag London 2012

The coordination challenges are functions of the complexity of processes, technology and also of the number and size of the different units. The degree of autonomy and spatial and cultural distance are explanatory factors of coordination challenges. In this picture the outsourcing decisions appear particularly important, and coordination mechanisms and other infrastructural issues need to be an integrated part of the outsourcing decision.

14.2 Agency Theory

Coordination is a wide field of research, ranging from rational shop floor issues up to, for example, socio-cultural issues and political systems. Organizational theorists have been particularly important contributors to the field of coordination mechanisms. For example, Mintzberg [1, p. 113] defines coordination mechanisms as: "the regulations used to coordinate work-related processes and orient individual activities towards the aims of the organization".

Agency theory has often been used in discussions about coordination. Key elements in this theory are opportunistic hyper rational actors with asymmetrical information searching for optimal incentive schemes to regulate their transactions. In more recent literature, agency theorists have included many agents, many principals or multiple tasks, and some have interpreted the models to illustrate implicit rather than explicit contracts [2]. However, the rationality might be 'disturbed', as decision processes are ambiguous [3], social norms can arise by accident [4], and organizations that struggle for survival [5], legitimacy [6] and power [7] can impact on rationality.

Thompson [8] hypothesized three patterns of dependency among organizations with corresponding coordination mechanisms: pooled, sequential, and reciprocal. He also suggested organizational hierarchies that are in cluster groups, with reciprocal interdependencies, sequential interdependencies, and finally those with pooled interdependencies. Van de Ven et al. [9] focused on three modes of coordination work activities: impersonal (plans and rules), personal (vertical supervision), and group (formal and informal meetings). Based on Thompson's [8] view of dependency, Van de Ven et al. [9] added a fourth mode, team arrangement, in which tasks are worked on jointly and simultaneously.

14.3 Mechanisms for Coordination

Galbraith [10] adopted an information processing viewpoint in relation to organization, suggesting the following mechanisms: lateral relations, direct contact, liaison roles, task forces, teams, integrators, integrating departments, and matrix organizations. Mintzberg [11] associates coordination mechanisms to organizational factors such as specialization, decentralization, formalization, size, age (maturity), environment, and power. He focuses on the means of coordination mechanisms, but emphasizes that "they are as much concerned with control and communication as with coordination" [11, p. 4]:

- *Mutual adjustment*—in a process of informal communication between people conducting interdependent work
- *Direct supervision*—one individual taking responsibility for the work of others
- *Standardization of work processes*—coordination occurs before an activity is undertaken
- *Standardization of output*—communication and clarification of expected results, actions required to obtain a goal, are not prescribed
- *Standardization of skills and knowledge*—people are trained to know what to expect of each other and to coordinate in an almost automatic manner
- *Standardization of norms*—socialization is used to establish common values and beliefs in order for people to work toward common expectations.

Jacobsson and Larsson [12] emphasize that organizational culture leads to the direct adjustment of members through a common base for action with shared understanding, expectations, and direction of the co-workers. Furthermore, socialization and maintenance mechanisms 'built into' a culture, will coordinate the frames of reference of the individuals and thereby indirectly coordinate their actions. This indirect coordination could be part of the inherent principles of a manufacturing paradigm, such as lean manufacturing, where teamwork and continual improvement become cornerstones [13].

Reger [14] groups coordination mechanisms into four categories:

- *Structure and formal mechanisms*—including centralization and decentralization of the decision-making process structure coordination bodies, programming and standardization, planning, and the control of results and behavior
- *Informal mechanisms*—focusing on company culture and informal aspects in coordination, integration, and control, but unlike the structure and formal instruments of coordination, these are not mechanisms with which coordination or its effects can be precisely planned or controlled
- *Hybrid coordination mechanisms*—includes combinations of the above types of mechanisms
- *Internal quasi-markets and internal prices*—normally based in agency theory and assumes that there are internal markets as a driving force, for example, for knowledge.

14.4 Coordination of Innovations and Knowledge Creation

Figure 14.1 shows Reger's [14] mechanisms for the coordination of R&D activities categorized in four groups; "formal/structural" such as planning and budgeting, decision procedures, etc.; "Hybrid/overlying" for example task forces and strategic

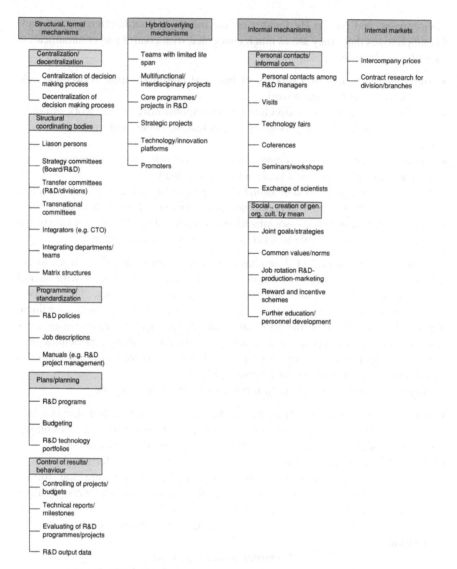

Fig. 14.1 Categories of R&D coordination mechanisms [14]

projects; "informal" such as common goals, visits/job rotation, seminars, etc.; and "internal markets" which are different types of internal services and contracts with basis in agent theory.

These coordination mechanisms are all relevant within manufacturing, in particular R&D activities. However, the table also presents means and enablers for coordination at tactical and operational levels for example in continuous improvement. An outsourcing partner would normally need more than the formal, contractual motivation for collaboration in innovation and knowledge transfer. Personal contacts, culture and different types of carrots and sticks are also important coordination mechanisms.

Previous research has uncovered a number of factors with impact on knowledge transfer between units in global companies. These factors would often represent barriers to knowledge transfer [15]:

- A probable trade-off between resources deployed in knowledge development and resources deployed in transfer to other units [16, 17]
- Conflicts of interest are likely to emerge between units [18–20]
- The character of the knowledge itself, its senders and recipients, and the relationship between them [21–23].

14.5 Implementing Outsourcing: Global Projects

When a decision regarding outsourcing has been reached, a plan for implementation is needed. There are developments to be carried out, negotiations with local authorities, buildings to be constructed or rented, equipment to be procured, commissioned and implemented, and an organization to be developed.

The mother organization is normally the driver of this process, and most commonly it is organized as a project with a project manager and a local project organization.

Project management is a well-developed profession with proven methods for planning and controlling projects. That includes complex construction projects. The internationally accepted procedure for managing projects is laid down in standards such Project Management Institute (PMI) Body of Knowledge [24]. This standard declares nine knowledge areas with a number of processes that represents good practice for most of the projects most of the time. The nine knowledge areas are:

1. Project integration management
2. Project scope management
3. Project time management
4. Project cost management
5. Project quality management
6. Project human resource management
7. Project communication management
8. Project risk management
9. Project procurement management.

A project is defined as "temporary endeavour undertaken to create a unique product, service or result" [24]. Managing a delivery or construction project requires management of the three main control parameters of a project: scope, time and cost [25]. There are commercial tools available for planning and controlling projects. It is beyond the scope of this book to go further into that.

Projects are managed to reach objectives that are defined before the project starts. Normally there are three different sets of objectives [25]:

- Projects objectives that define what the project manager should deliver at the end of the project
- Business objectives that define what the owner (company that outsources) expects to obtain with the results of the projects (i.e. operation of the new plant)
- Social objectives that define what the society expects of results from the project.

The social objectives are of importance for an outsourcing project. The local society where the plant is constructed may have expectations as to job opportunities, tax revenues, supply of goods and services from the local community, etc. Quite often outsourcing companies are expected to render some services to the local community, for example in the form of infrastructure or even sponsoring of activity in the local community (corporate social responsibility).

The project execution is influenced by a number of stakeholders that may have different agendas. There are three main groups of stakeholders [25]:

- Project owner (outsourcing company)
- Project organization (those involved in executing the project from the outsourcing company and all contractors that delivers goods and services)
- Society (local authorities, service organizations, interest organizations and media).

As these stakeholders may influence the project in a way that is not always predictable, they introduce risk into the project. In all major projects, risk management is consequently an important task to address. Outsourcing projects are almost always global which also introduces a number of frame conditions and risks that may not exist in a traditional national project. This adds to the complexity and challenges the outsourcing company striving for a successful project.

There is a difference between global project business and large global projects. Global projects business is undertaken by companies where their main business is developing projects around the globe. This may be the case for the oil industry and international engineering and consulting companies. However, this is rarely the situation for a company outsourcing some of its production. They are executing a single or a few global projects. The point is that the outsourcing company has not had the opportunity to develop the special competence needed for global projects to same degree as those companies that are in global project business. This makes it even more challenging for the outsourcing company that in addition must succeed to maintain its competitive advantage.

The added complexity in global projects is mainly due to:

- Differences in culture
- Communication due to different time zones
- Geographical distance
- New stakeholders that are unfamiliar to the project organization.

The most important aspect is probably the difference in culture. Many may have problems adapting to the local culture, understanding how business is done and how people react. The classic work on cultural understanding is a major research carried out by Hofstede. Hofstede [26] originally defined four dimensions of cultural consequences:

- Power distance
- Masculinity versus femininity
- Individualism versus collectivism
- Uncertainty avoidance.

An index per country is defined for each of these dimensions. The power distance index expresses the extent to which the less powerful members of institutions and organizations within a country expect and accepts that power is distributed unequally. The masculinity index expresses the gap between gender roles, i.e. degree of difference between 'masculinity' and 'femininity'. The individualism index expresses the looseness of ties between individuals. It is the opposite of collectivism—the degree to which individuals are integrated into groups. The uncertainty avoidance index expresses the tolerance for uncertainty and ambiguity.

Hofstede later added a fifth dimension: long- versus short-term orientation. His work has been criticized because the data that the analysis builds on stems from the 1970s and may be perceived as outdated. However, most researchers refer to Hofstede as one of the main contributions in the field.

References

1. Mintzberg H (1991) Mintzberg über management. führung und organization. Mythos und realität. Gabler, Wiesbaden
2. Eisenhardt K (1989) Agency theory: an assessment and review. Acad Manage Rev 14(1):57–74
3. March JG, Olsen JP (1976) Ambiguity and choice in organizations. Universitetsforlaget, Bergen
4. Elster J (1989) Social norms and economic theory. J Econ Perspect 3(4):99–117
5. Nelson RR, Winter SG (1982) An evolutionary theory of economic change. Belknap Press of Harvard University Press, Cambridge
6. Parsons T (1960) Structure and process in modern societies. Free Press, Glencoe
7. Pfeffer J (1981) Power in organizations. Pitman, Marshfield
8. Thompson JD (1967) Organizations in actions: social science bases of administrative theory. McGraw-Hill Book Company, New York
9. Van de Ven AH, Delbecq AL, Koenig R Jr (1976) Determinants of coordination modes within organizations. Am Sociol Rev 41(4):322–338
10. Galbraith JR (1973) Designing complex organizations. Addison-Wesley, Reading
11. Mintzberg H (1979) The structuring of organizations. Prentice-Hall, Englewood Cliffs
12. Jacobsson T, Larsson R (1984) Organizational culture and coordination of action: intersubjective frames of reference in action. In: Proceeding of the first international conference on organizational symbolism and corporate culture, University of Lund, Sweden
13. Womack JP, Jones DT, Roos D (1990) The machine that changed the world: the story of lean production. Harper Business, New York

14. Reger G (2004) Coordinating globally dispersed research centres of excellence: the case of Philips electronics. J Int Manage 10(1):51–76
15. Bjorkman I, Barner-Rasmussen W, Li L (2004) Managing knowledge transfer in MNCS: the impact of headquarters control mechanisms. J Int Bus Stud 35:443–455
16. Forsgren M, Johansson J, Sharma D (2000) Development of MNC centres of excellence. In: Holm U, Pedersen T (eds) The emergence and impact of MNC centres of excellence. Macmillan, London, pp 45–67
17. Szulanski G (1996) Exploring internal stickiness: impediments to the transfer of best practice within the firm. Strateg Manage J 17:27–43 (Winter Special Issue)
18. Birkinshaw J, Hood N (1998) Multinational subsidiary evolution: capability and charter change in foreign-owned subsidiary companies. Acad Manag Rev 23(4):773–795
19. Levitt B, March JG (1988) Organizational learning. Annu Rev Sociol 14:319–340
20. Forsgren M (1997) The advantage paradox of the multi-national corporation In: Bjorkman I, Forsgren M (eds) The nature of the international firm: nordic contributions to international business research. Copenhagen Business School Press, Copenhagen DK, pp 69–85
21. Zander U, Kogut B (1995) Knowledge and the speed of the transfer and imitation of organizational capabilities: an empirical test. Organ Sci 6(1):76–92
22. Nonaka I, Takeuchi H (1995) The knowledge creating company. Oxford University Press, New York
23. Hansen MT (1999) The search-transfer problem: the role of weak ties in sharing knowledge across organization subunits. Adm Sci Q 4(1):82–111
24. PMI (2008) A guide to the project management body of knowledge, 4th edn. Project Management Institute, Newton Square
25. Rolstadås A (2008) Applied project management—how to organize, plan, and control projects. Tapir academic press, Trondheim
26. Hofstede G (2001) Culture's consequences: comparing values, behaviors, institutions, and organizations across nations. Sage, Thousand Okas

Part IV
Innovation and Knowledge Transfer

Innovation and knowledge are terms that are closely related. As we have seen manufacturing companies need to innovate to survive over time. These processes include not only different parts of the company but also partners in the production. In this part of the book knowledge aspects related to strategic decisions are described and exemplified. Different manufacturing paradigms represent different prerequisites and challenges for the knowledge creation process due to the innovation process, employees involvements, and so on. Challenges and how they might be dealt with are illustrated through practical examples.

Chapter 15
The Innovation Process

Abstract Continuous improvement and incremental change is not enough—companies also need to be part of major changes or radical and disruptive innovation. Innovation is necessary in all organizations in order to maintain or improve competitive position. Innovation is no less important in manufacturing than in any other section of industry. The 'Oslo Manual' defines innovation as, "… the implementation of a new or significantly improved product (good or service), or process, a new marketing method, or a new organizational method in business practices, workplace organization or external relations". Radical innovation is about making major change and we can visualise it as a 'step change' in some measure of performance such as revenue or efficiency. A high level of risk is not possible for many companies who prefer instead to invest in incremental innovation or continuous small changes to their products, processes and services. Incremental innovation is supported in manufacturing by initiatives such as the Plan Do Check Act cycle and 'Lean Manufacturing'. Most manufacturing organisations need to consider both radical and incremental innovations. Manufacturing paradigms have inherent principles and guidelines for the development of innovation processes. Projects and project models are essential for building an efficient innovation 20 process.

15.1 What is an Innovation

In an ever-changing economic landscape, innovation has become a key competitive factor in manufacturing. Kevin Kelly, founding executive editor of Wired magazine, states that "wealth today flows directly from innovation, not optimization. It is not gained by perfecting the known, but by imperfectly seizing the unknown" [1]. Continuous improvement and incremental change is not enough—companies also need to be part of major changes. Innovation is no less

A. Rolstadås et al., *Manufacturing Outsourcing*,
DOI: 10.1007/978-1-4471-2954-7_15, © Springer-Verlag London 2012

Fig. 15.1 Radical and
incremental innovation

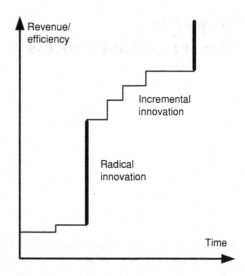

important in manufacturing than any other section of industry. Apple Corporations
most innovative products for example were successful only because they were
coupled with highly innovative manufacturing and distribution processes that kept
costs low and reliability and throughputs high.

Innovation has been defined as the process of making changes in something
established by introducing something new [2]. Innovation is necessary in most
organizations in order to maintain or improve competitive position. Schumpeter
[3, p. 65] defines innovation as, 'combinations' of new or existing knowledge and
resources. He describes five innovation areas:

- *Product innovation*—introduction of a new good or a new quality of a good
- *Process innovation*—introduction of a new method of production
- *Market innovation*—opening of new markets
- *Input innovation*—conquest of a new source of supply of raw material
- *Organizational innovation*—development of a more effective organization.

The 'Oslo Manual' [4] is one of the foremost international sources of guidelines
for the collection and use of data on innovation activities in industry. In it,
innovation is defined as:

> … the implementation of a new or significantly improved product (good or service), or
> process, a new marketing method, or a new organizational method in business practices,
> workplace organization or external relations.

This definition of innovation makes no reference to the size and scope of the
change. Innovation in manufacturing can be either radical or incremental. Radical
innovation is about making major change [5] and we can visualise it as a 'step
change' in some measure of performance such as revenue or efficiency, see
Fig. 15.1. Most organizations engage in some form of radical innovation over their
lifetime; however, it is also often resource intensive and risky.

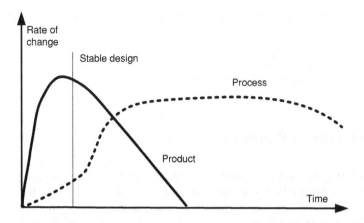

Fig. 15.2 Product and process innovation

A high level of risk is not possible for many companies who prefer instead to invest in incremental innovation or continuous small changes to their products, processes and services. Incremental innovation is supported in manufacturing by initiatives such as 'Continuous Improvement', 'Lean Manufacturing', 'Six Sigma', 'Total Quality Management' and 'World Class Manufacturing'.

Most organizations adopt a dual approach to the size and scope of their innovation activities by adopting incremental innovations that will yield results in the short term, while also having some radical innovations that may yield significant results in the long term. Every now and then a radical innovation is introduced that disrupts business practice [5] and in our discussion the practice of how to perform the manufacturing process in a more radical way. These disruptive innovations often occur through the deployment of new sciences and technologies. Arguably the largest disruptive technology to emerge recently in manufacturing has been the World Wide Web that has disrupted the way organizations manage their supply chains and many aspects of customer relationship management and enterprise resource management. Most companies 'watch out' for new technologies, for example a new types of wireless sensors, as they emerge from research laboratories in order to assess their competitive advantage in driving costs lower and reliability higher.

Process innovation is very closely associated with product innovation. Every new product needs to be manufactured and processed before it reaches the customer. If the process cannot produce the right level of cost, quality and reliability then any new product can be rendered useless. Most product innovation takes place at the beginning of the products life. Initial product ideas are tested and refined and final changes are made following pilot production. The product reaches a state of 'stable design' and any changes to the product after production begins are fewer, less frequent and more costly. See Fig. 15.2.

Process innovation on the other hand typically only begins after the product has reached stable design. Innovations to the manufacturing process then occur throughout the life of the product. Manufacturers often look for changes to the process on an annual basis—changes that will reduce costs, increase quality or reduce lead times i.e. cost reduction programmes.

15.2 Innovation Processes

Manufacturing has inherent principles and guidelines for how 'things get done' and these are often reflected in the innovation process. Lundvall [6] distinguishes between interactive and linear models of innovation. A linear model has well-defined sequences and tasks and will often rely on analytical knowledge. The interactive model involves an interactive learning process on a task between practitioners and experts (e.g., an R&D department). The interactive model will normally rely on both a synthetic and an analytical knowledge base. The innovation process is more decentralized, interactive and incremental within lean manufacturing than in mass manufacturing [7]. More people are involved in the lean knowledge creation process than in the more linear, formalized, and centralized innovation processes of mass manufacturing.

Innovation is essentially the process of generating ideas that lead to change. Change in manufacturing is the process of converting an organization or any part of it, from its current state to some future desired state. Change can be planned or emergent. *Planned change* is a formal and typically annual process of converting the organization from one state to another. A programme or plan is created which identifies what needs to be changed, by whom and when. The present state of the organization is usually articulated in terms of current products, processes, structures and so on (Fig. 15.3). The future state of the organization is articulated through entirely different terms such as its goals, ideas and projects that when implemented create the new organization.

Emergent change is contingent on other changes taking place in the environment and recognises that change is an open ended process. It involves an open management style and greater empowerment of individuals for taking the decisions necessary for change. It also recognises the need for experimentation and adaptation to change. A difficulty with the *emergent approach* is that individuals sometimes lose the urgency to create change. Both approaches—planned and emergent—have advantages and disadvantages. A hybrid approach to change and innovation adopts the better features of both. Planned change can be useful for creating the annual sense of urgency whereas the open management style can empower individuals to make changes to plans as the needs emerge.

There are a broad variety of methods of creating plans that manufacturing organizations can use to facilitate the broader steps in the innovation process. Methodologies are an important step by step approach to achieving a set of end goals.

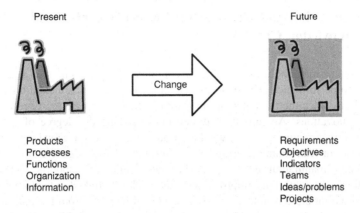

Present   Future

Products   Requirements
Processes   Objectives
Functions   Indicators
Organization   Teams
Information   Ideas/problems
                                                 Projects

Fig. 15.3 Present and future organization

One method of note has been promoted by the Harvard's John Kotter [8]. His method outlines eight steps:

1. Establishing a "sense of urgency"
2. Forming a "powerful guiding coalition"
3. Creating a vision
4. Communicating the vision
5. Removing obstacles for "acting on the vision"
6. Planning for and "creating short-term wins"
7. Consolidating Improvements
8. Institutionalizing "new approaches".

Establishing urgency involves looking hard at organizations' competitive position, communicating this information broadly and dramatically, motivating staff and employees, looking for leaders and champions of change, and discussing unpleasant facts openly. Forming a coalition involves developing a strong bond and loyalty between managers and the company. Creating a vision involves developing a mental image of a possible and desirable future state of the manu-facturing system that is realistic, credible and attractive, and that most individuals can buy in to. Communicating the vision involves winning the hearts and minds of individuals. All existing and many new communication channels should be utilised and clearly behaviour must match words. Removing obstacles involves identifying resistance to change as early as possible. Creating and achieving short term wins is necessary for boosting morale and convincing everyone that overall success is possible. Consolidating improvements involves making sure that changes sticks and that things don't return to the old norms. Finally, institutionalising new approaches involves making sure that innovations lead to lasting change within the organization and achieve their full potential.

15.3 Innovation and Centralized Versus Decentralized Knowledge Creation

Innovation in manufacturing is often a combination of incremental and radical changes to performance. There is often interplay between incremental and radical innovations, since the incremental innovations often illuminate the need for more radical innovations. An important dimension regarding the degree of change is whether a manufacturing strategy is based on a decentralized knowledge creation process, as in lean manufacturing, or a centralized knowledge creation process, as in mass manufacturing. The decentralized approach supports incremental innovations and continuous improvement, and is closely linked to tacit knowledge. In this situation there is a need to extend cycles of learning from the individual or team level to the organizational level and even in some cases the whole production system. A challenge is to transfer tacit knowledge to other parts of the company and production system.

In turbulent times it is easy to give up on innovation. The uncertainty associated with success rates for new ideas and the challenges in implementation and industrialization leaves many with the sense that innovation—the creation of new value—is mysterious, unpredictable and apparently, unmanageable. Searching for breakthroughs is expensive and time consuming, and many managers fall back on incremental improvements to existing products and services.

A centralized knowledge creation approach is more relevant when the innovations are more radical and the knowledge more explicit. The knowledge transfer challenges will differ as tacit knowledge has to be made more explicit in order to be transferred to other units in the lean company. Transfer challenges related to explicit knowledge will be to make the knowledge relevant in a specific context or working environment.

15.4 Coordinating Innovation Processes: Project Models

Innovation activities are not necessarily the same as innovation, but rather enablers or assumptions for innovations. Innovation activities covers are all scientific, technological, organizational, financial and commercial steps which actually, or are intended to, lead to the implementation of innovations.

Some innovation activities are themselves innovative; others are not novel activities but are necessary for the implementation of innovations. Innovation activities also include R&D that is not directly related to the development of a specific innovation [4]. All these kinds of activities could be part of an innovation or innovation process. This implies coordination and management challenges.

The coordination approaches and mechanisms relevant are those presented in Chap. 14 for example the agent theory, and are closely related to knowledge creation and knowledge transfer. However, there is also an important element of

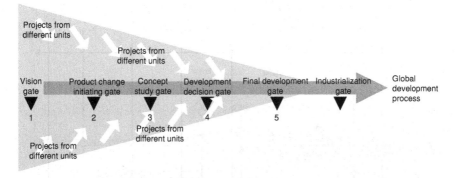

Fig. 15.4 The product development model for an OEM in the automotive industry [7]

creativity in the innovation process and coordination of innovation. A challenge is then to balance the creativity element in innovation with the need for structure and planned actions. Innovations are often organized through and many project models have been designed to improve the innovation processes. The challenges could be even greater to initiate and capture innovativeness and improvements in the daily operations and activities. For a company to keep up the innovative pace from for example shop floor activities they need a structure for launching and discuss, verify, implement ideas. This might include arenas like 'stand up meetings', 'team leader meetings', and so on, but also often a defined format for the ideas, a way of describing them and make them more explicit.[1]

The project model in Fig. 15.4 is an example from a major OEM producing trucks, but similar to what we could find in most industries for R&D projects. The figure illustrates how different projects are entering into the R&D pipeline at different stages, but where the gates are reducing the number as they are moving towards implementation and industrialization. In a decentralized manufacturing structure, for example an outsourced structure, several of the decision gates could be at contractor level, especially those not involving big funding (gates 1–3). However, the overall objectives and strategies should provide the premise for continued development consideration. These premises in turn require tools and methodologies that could give us some quantified analyzes.

In Fig. 15.5 the project model of one of the suppliers to the truck OEM is presented. The project development model is quite similar (and is integrated) to the one of the OEM but represents different concepts that the supplier launches or presents to the OEM. The OEM does not necessarily accept any of the concepts developed by the supplier. However, in lean manufacturing the supplier and OEM have particular integrated development and improvement agendas. This is in particular the case for the most strategic suppliers (first tier). The knowledge creation model implies a particular focus on the initial concept phases, up to phase

---

[1] Such a structure is described in the case section.

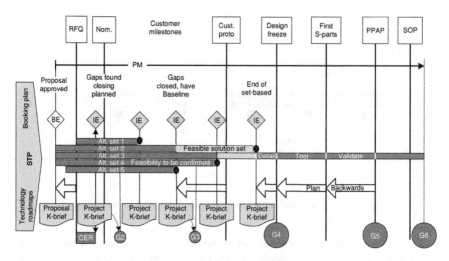

Fig. 15.5 Product development model for a supplier in the automotive industry

gate 3 (G3) of the project development model. In these phases, people from different fields concurrently describe concepts, and feasibility and business effects are evaluated.

Industrial structures vary in their ability to coordinate information flows necessary for innovation and to overcome power relationships adverse to innovation. A central debate in industrial policy today is that between proponents of large vertically integrated firms on the one hand and those of networks of small specialized producers on the other [9]. Institutional alternatives are many, both for companies and networks. The relative desirability of the various structures, then, depends on the nature and scope of technological change in the industry and on the effects of various product life-cycle patterns.

15.5 Managing Innovation

Each year manufacturing organizations budget a percentage of turnover on innovation i.e. making changes to their established processes and services. The amount of investment can vary significantly between organizations. The average investment across all types of organizations is 4% [11]. This budget will typically be spread across various functions including marketing, product design, information systems and manufacturing systems. The principle goals required in return for this investment varies between organizations. The following have been found across a large number of manufacturing and services organizations and ranked in order of popularity:

- Improved quality
- Creation of new markets
- Extension of the product range
- Reduced labour costs
- Improved production processes
- Reduced materials
- Reduce environmental damage
- Replacement of products/services
- Reduced energy consumption and
- Conformance to regulations.

From the above list 'improving quality' is the most popular goal for organizations to invest in innovation. Quality relates to product and process reliability and function. Attaining goals is the ultimate objective of an innovation process but unfortunately, most innovation efforts fail to varying degrees. Failure is common and failure rates vary between organizations and across industrial sectors. Some research quotes failure rates of fifty percent while other research quotes failure as high as 90% [12]. Failure is an evitable part of the innovation process and most successful organizations factor in an appropriate level of risk for not achieving their goals. The impact of failure goes beyond the simple loss of investment. Failure can also lead to loss of morale among employees and an increase in cynicism and even higher resistance to change in the future. On the other hand failure is an important learning opportunity, when honestly observed and analyzed. Early screening of potential failures avoids unnecessary resource allocation and an earlier decision to shift focus towards potential successes.

Common causes of failure within the innovation process in most organizations can be distilled into five areas where organizations need to give more attention [13]:

- Better definition of goals
- More effective alignment between actions and goals
- Greater participation of individuals in teams
- Better monitoring of results
- Great communications and sense of community.

Goals: A goal is the defined as "the objective of an effort". There are a number of ways of defining goals, these include: (1) Statements such as the mission and vision statement; (2) Requirements of stakeholders such as customers and shareholders; (3) Objectives such as strategic plans; (4) Standards such as ISO (International Organization for Standards) 9,000 and Safety regulations and (5) Indicators of performance such as output and profits.

Actions: An action is defined as "the expenditure of effort". Actions include such activities as: (1) Problem identification; (2) Idea generation; (3) Managing initiatives and projects and; (4) managing project portfolios. A key issue is that actions are in some way connected with the goals of the organization.

Teams: A team is the defined as "resources for an effort". Teams are made up of individuals and there are a number of issues related to greater participation by individuals in teams. These include: (1) assigning responsibility; (2) building structure in teams; (3) improving participation by individuals; (4) linking the performance of individuals to organizational goals and; (5) appraising performance of individuals.

Results: The term result is defined as "the outcome of an effort". The principal results that an organization needs to concern itself with are the results and progress of goals, actions and teams. Results analysis also involves reviewing connections between for example goals and actions or teams and goals.

Communities: Community are defined as "individuals with a common purpose". That common purpose is the goals of the organization but may also reflect the professional goals of the individuals in it. Building community is time consuming and resource intensive process involving many key issues such as leadership, communications and knowledge management.

The innovation funnel provides a useful representation for the information requirements for managing goals, actions, teams and results. The funnel illustrates how goals, actions, teams and results interact with each other to create change in a manufacturing organization—see Fig. 15.6. The funnel contains four arrows. Each arrow represents the flow of goals, actions, teams and results. Actions enter the wide mouth of the funnel and represent among other things, alternative ideas for change. These actions flow towards to the neck of the funnel where many will be eliminated. The neck of the funnel is constrained by two arrows—goals and teams. These constraints loosen or tighten the neck of the funnel. Tightly defined goals close the neck of the funnel allowing fewer ideas through. The availability of more teams or skills on the other hand opens neck of the funnel allowing more ideas to flow through. The final arrow, 'results', flows from the narrow end of the funnel and represents information concerning the results and progress of goals, actions and teams. This arrow flows back towards goals representing the impact that results and learning have on the process of defining and redefining goals.

An important aspect of the innovation funnel is the connections generated between goals, actions and teams. Ideas, for example, that cannot easily be connected to a goal will find it difficult to proceed into the funnel. This has two affects; firstly the teams generating the idea will study the goals more closely and perhaps change their ideas and secondly; good ideas will begin to impact on the ways goals have been defined or will be redefined. The same analysis is true for the connections between say ideas and team or more particularly the skills of teams. This is a natural learning process within an innovation community. When specific goals change there is a knock on effect in the generation of ideas from others in the organization. This learning process gives the innovation community the ability to change the innovation process in response to changing demands of stakeholders and emergent trends in technologies and other enablers of innovation.

The funnel represents a 'portfolio' of actions constrained or guided by specific goals and teams. As an action progresses forward it changes from starting as a

Fig. 15.6 Innovation funnel

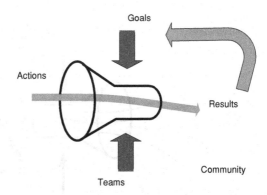

problem, to becoming an idea and emerging as an approved project. It emerges from the funnel process and enters a 'project management' phase—See Fig. 15.7. There are a number of ways in which the progress of a project can be monitored and managed after it emerges from the innovation funnel. The specific project can be designed to pass through various 'stage gates'[14] where each gate represents a decision point on whether to adjust various project parameters or cancel the project altogether due to changing circumstances. Simultaneously, the decision around changing a project can involve the performance of the entire portfolio of projects as a whole. Perfectly good projects may be adjusted or even postponed because of the demands of the portfolio at large.

15.6 Innovation and R&D Models

Edward Deming has been widely accepted as one of the world's most prominent authorities on quality management. He gained credibility from his influence on Japanese and American industry [15]. Although he was best known for his emphasis on the management of a system for improving quality, his thinking was based on the use of statistics for continual improvement. His mantra was "continual never-ending improvement" and poor quality was not the fault of labour, but a resulted of poor management of the system for continual improvement.

Even if the improvement cycle (Shewhart cycle) was initially developed and described as an approach for continuous improvement, it also represents a framework for R&D projects and strategy development. Deming's model was based on several iterations around four "PDCA" steps (Fig. 15.8):

- *Plan*—establish the objectives and processes necessary to deliver the expected results
- *Do*—implement the new processes, preferable on a small scale
- *Check*—measure the new processes and compare the results against the expected results to ascertain any differences

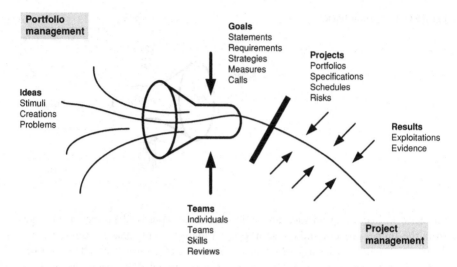

Fig. 15.7 Portfolio and project management

- *Act*—analyze the differences to determine causes. Each will be part of either one or more of the PDCA steps and determine where to apply changes that will include improvement.

When a pass through these four steps does not result in the need to improve, refine the scope to which PDCA is applied until there is a plan that involves improvement. The PDCA approach is much used in incremental innovations where innovations are made stepwise and often refined through iterations.

Improvements have to take place continuously within the whole supply chain. These improvement processes need to be linked as illustrated in Fig. 15.9 to get the best effect. This linkage is a question of knowledge transfer and information flow within and across organizational boundaries.

A development model that has received increased attention as a framework for innovation is the 'spiral approach' [17]. This approach captures key aspects of incremental innovations since innovations are made through several iterations and refinements from an initial basic idea. Then as learning progresses, more and more details are introduced, while at the same time they are related to the basic PDCA steps.

The 'waterfall model' is often used in large development projects and in radical innovations. This is a sequential design process, in which progress is seen as flowing steadily downwards like a waterfall through the phases of 'conception', 'initiation', 'analysis', 'design', 'construction', 'testing and Maintenance'. The waterfall development model originates in the manufacturing and construction industries: highly structured physical environments in which after-the-fact changes are prohibitively costly, if not impossible. Using the PDCA as reference, the waterfall model maintains that one should move to a phase only when it's

Fig. 15.8 The Shewhart-/
PDCA cycle [16]

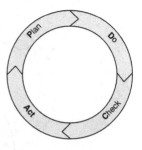

proceeding phase is completed and perfected. This means that there is an objective to reduce the iterations. However, there are various modified waterfall models, including Royce's [18] final model that includes slight or major variations to this process.

15.7 PDCA in Incremental and Radical Innovation

Radical innovations often comprise entirely new products, and are often undertaken by new entrants with diversified knowledge. Minor improvements in existing products and processes constitute incremental innovations, often undertaken by incumbent firms with a specific knowledge base. However, incremental innovations might represent world class innovations over time, when organized as many small steps in contrast to the few large innovative steps in radical innovation.

As we see from Fig. 15.10 the innovations in companies are often a combination of incremental and radical innovations. The importance and innovation degree are often as high for incremental as for radical innovations, but where incremental innovations are going through more PDCA cycle iterations and hence need more time. There is also often interplay between the incremental and radical innovations since the incremental innovations often enlightens the need for more radical innovations. Hence incremental innovations often trigger the more radical.

In manufacturing, innovation is frequently incremental and continuous through, for example, quality improvement and various lean approaches, but it can also be radical and disruptive [19]. An important dimension is whether a manufacturing strategy is based on a decentralized knowledge creation process, as in lean manufacturing, or a centralized knowledge creation process, as in mass manufacturing. The decentralized approach supports incremental innovations and continuous improvement, and is closely linked to tacit knowledge.

In a decentralized structure, and where the knowledge creation process is incremental, involving people at shop floor level. In this situation there is a need to extend cycles of learning from the individual or team level to the organizational level and even in some cases to the whole production system. A challenge is then to transfer tacit knowledge to other parts of the company. Radical innovations are often R&D projects involving many resources from outside the company.

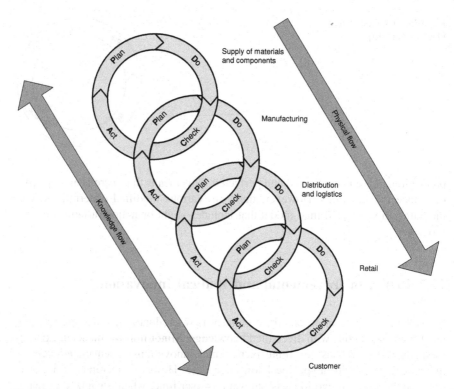

Fig. 15.9 PDCA driving innovation processes in supply chains

The risks of not succeeding in reaching breakthrough objectives such as new technologies in products and/or processes often require advanced project models focused on reducing uncertainty (risk). Decision gates are normally formalized though the use of evaluation decision criteria and methodologies.

In turbulent times it is easy to lose focus on creating more breakthrough innovation and focus instead on maintaining the status quo. The uncertainty associated with success rates for new ideas and the challenges in implementation leave many with the sense that innovation—the creation of new value—is mysterious, unpredictable and apparently, unmanageable. Searching for breakthroughs is expensive and time consuming, and many managers fall back on incremental improvements to existing products and services.

A centralized knowledge creation approach is more relevant when the innovations are more radical and the knowledge more explicit. The knowledge transfer challenges will differ as tacit knowledge has to be made more explicit in order to be transferred to other units in the lean company. Transfer challenges related to explicit knowledge will include making the knowledge relevant in a specific context or working environment.

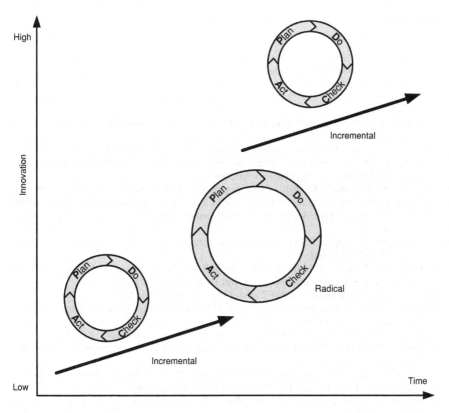

Fig. 15.10 Incremental versus radical innovations and the PDCA Cycle

15.8 Manufacturing Paradigms and Innovation

Manufacturing paradigms have inherent principles and guidelines for how 'things are done', problem solving, stakeholder relations, and so on. This is typically reflected in the innovation process adopted by a particular organization. Lundvall [20] distinguishes between "interactive" and "linear" models of innovation. A linear model has well-defined sequences and tasks within, for example, R&D projects. The linear model will often rely on analytical knowledge, while the interactive model is more relevant in incremental innovations. An interactive learning process between practitioners and experts (e.g., an R&D department) will normally rely on both a synthetic and an analytical knowledge base.

The innovation process is much more decentralized, interactive and incremental within lean manufacturing than in mass manufacturing [14]. More people from different backgrounds and skill competencies are involved in the lean knowledge creation process than in the more linear, formalized, and centralized innovation processes we see in mass manufacturing.

References

1. Kelly K (1998) New rules for the new economy. Fourth estate, London
2. Fagerberg J, Mowery DC, Nelson RR (2005) The Oxford handbook of innovation. Oxford University Press, London
3. Schumpeter J (1934) The theory of economic development. Harvard University Press, Cambridge
4. OECD (2005) Oslo manual: guidelines for collecting and interpreting innovation data, 3rd edn. OECD, Paris
5. Christensen CM (1997) The innovator's dilemma. Harvard Business School Press, Boston
6. Lundvall BÅ (1992) National systems of innovation: towards a theory of innovation and interactive learning. Pinter Publishers, London
7. Henriksen B, Røstad CC (2010) Evaluating and prioritizing projects—setting targets. The business effect evaluation methodology BEEM. Int J Managing Projects Bus 3(2):275–291
8. Kotter JP (1990) A force for change: how leadership differs from management. Free Press, New York
9. Robertson PL, Langlois RN (1995) Innovation, networks, and vertical integration. Res Policy 24(4):543–562 (Elsevier)
10. Gerybadze A, Reger G (1999) Globalization of R&D: recent changes in the management of innovation in transnational corporations. Res Policy 28:251–274
11. http://cordis.europa.eu/itt/itt-en/99-7/innov3.htm. Accessed June 2006
12. Strebel P (1999) Why innovation fails, in Harvard business review on change. Harvard Business Press, Boston, pp 139–157
13. O'Sullivan D, Dooley L (2010) Applying innovation. Sage Publications, Los Angeles
14. Cooper RG (2000) Product leadership: creating and launching superior new products. Perseus, New York
15. Foster TS (2006) Managing quality—Integrating supply chain. Prentice Hall, New Jersey
16. Deming WE (1986) Out of crisis Boston. MIT/CAES, Cambridge
17. Boehm B (1988) A spiral Model of Software Development and Enhancement. Computer 21(5):61–72
18. Royce W (1970) Managing the development of large software systems. Proc IEEE WESCON 26:1–9
19. Tidd J, Bessant J, Pavitt K (2005) Managing innovation: Integrating technological, market and organizational change. Wiley, London
20. Henriksen B, Rolstadås A (2010) Knowledge and manufacturing strategy—how different manufacturing paradigms have different requirements to knowledge. Examples from the automotive industry. Int J Prod Res 48(8):2413–2430

Chapter 16
What is Knowledge?

Abstract Manufacturing is increasingly about knowledge creation, learning, flexibility and continuous improvement. Consequently, knowledge in manufacturing is not only a matter of systems and management. According to Bloom's taxonomy, the cognitive domain relates to mental skills (knowledge), the affective domain relates to growth in feelings or emotional areas (attitudes), while the psychomotor domain is concerned with manual or physical skills (skills). Knowledge management relates to business processes where the creation, transfer, and adaptation of knowledge are basic elements. A knowledge dimension that is similar to the 'general world' versus 'arbitrary specialized world', and that has been emphasized in the literature is the concept of 'tacit knowledge' versus 'explicit knowledge'. This dimension suggests that there are two types of knowledge: tacit, which is embedded in the human brain and cannot be expressed easily and explicit knowledge, which can be easily codified. Both types of knowledge are important, but Western organizations have focused largely on managing explicit knowledge.

16.1 Rationalism and Empiricism

Manufacturing is increasingly a question of knowledge, and hence cannot be organized and managed in the same way as in traditional manufacturing based on principles of scientific management and well-defined processes [1]. Manufacturing is now more about knowledge creation, learning, flexibility and continuous improvement. Consequently, knowledge in manufacturing is not only a matter of systems and management but equally about systems change.

This view questions the rationalism and empiricism approaches to knowledge that have dominated in manufacturing. "Rationalism," with roots in Greek philosophy [2], has not only dominated the Western philosophical tradition, but is still the

A. Rolstadås et al., *Manufacturing Outsourcing*,
DOI: 10.1007/978-1-4471-2954-7_16, © Springer-Verlag London 2012

dominating approach to knowledge within manufacturing. This approach assumes that any problem can be analyzed into basic elements and by explicit rules [3].

This focus on rationality—a structure to meet the knowledge requirements—is typically reflected in the work of Lin et al. [4], Wilson [5], and Filos and Banahan [6]. The rational approach is normally a discussion of structure issues and ICT systems, even if human aspects are also represented.

16.2 Pragmatic and Combined Views on Knowledge

The philosopher, economist and historian David Hume (1711–1776) claimed that human beings learn through association and that "truth springs from an argument among friends" [7]. This 'empiricist' approach to knowledge differs from rationalism as it views knowledge as coming from experience via the senses, and that science also flourishes through observation and experiment [7]. However, 'pragmatic' approaches arose in the late ninetieth and early twentieth century which accepted that knowledge could be attained both inductively from experience, and deductively by basic principles. The twentieth century philosophers Martin Heidegger and Ludwig Wittgenstein have been some of the most important contributors to this approach and their arguments for understanding the basic connections between "machination" and "lived-experience" emerged to one of the most important developments in the knowledge management [8]. Their view was that human understanding was 'knowing how' (i.e. finding one's way around the world), rather than on 'knowing what' (i.e. knowing facts and rules) [3].

Heidegger's and Wittgenstein's perspective has increasingly been adapted by the emerging socio-technical approach. Polyani [9] and Nonaka and Takeuchi [10] are proponents of pragmatic and combined views on knowledge related to manufacturing. There is now an increased understanding that knowledge, and the management of it, is influenced by culture, leadership, measurement, education, reward and incentive systems, organizational adaptability, values and norm, and technology [11]. These views about knowledge are also increasingly reflected in governmental policies and strategies such as the focus on the learning economy in the EU's Lisbon Agenda.

16.3 Knowledge and Information

Knowledge is embedded in people, and knowledge creation occurs in the interaction of people [12], and has a different meaning to the term 'information' [13]:

> Information is that commodity capable of yielding knowledge, and what information a
> signal carries is what we can learn from it.... Knowledge is identified with information-

produced (or sustained) belief, but the information a person receives is relative to what he or she already knows about the possibilities at the source. [13, p. 84–86]

Thus, knowledge can be viewed as the result of an interaction between intelligence (capacity to learn) and situation (opportunity to learn), and hence is socially constructed. Winterton et al. [14] make a distinction between knowledge that is 'general', essentially irrespective of any occupational context, and knowledge that is 'specific' to a sector or particular group of occupations.

16.4 Terms Related to Knowledge

The first and most influential generic typology of competence, skills and knowledge was developed by Bloom and colleagues in the 1960s for use in educational establishments [15–17]. Generally known as Bloom's taxonomy, it is based on three domains of educational activities: cognitive, affective, and psychomotor. The cognitive domain relates to mental skills (knowledge), the affective domain relates to growth in feelings or emotional areas (attitudes), while the psychomotor domain is concerned with manual or physical skills (skills).

Skill is normally regarded as a more 'practical, close to work' knowledge. Proctor and Dutta [18] emphasize four elements in their definition of skills:

- Develops over time, with practice
- Is goal-directed in response to some demand in the external environment
- Is acquired when components of behavior are structured into coherent patterns
- Cognitive demands are reduced as skill develops.

There is no uniform understanding of the term 'competence' [19, 20]. Competence is about describing the ability to perform a specific role and represent the combination of the three elements knowledge, skills and attitudes. These elements and the role they are related to define levels of competence. A behavioral definition of competence describes it as "a cluster of knowledge, skills, abilities, behaviors, and attitudes related to job success and failure" [21]. The PMI (Project Management Institute) [22, p. 2] defines competence as "a cluster of related knowledge, attitudes, skills, and other personal characteristics that:"

- "Affects a major part of one's job
- Correlates with performance on the job
- Can be measured against well-accepted standards
- Can be improved through training and development
- Can be broken down into dimensions of competence".

To be competent, one needs to be able to interpret the situation in a given context and to have a repertoire of possible actions, if relevant. Regardless of training, competence grows through experience and the extent of an individual's ability to learn and adapt the behavior element of competence.

Knowledge management is a part of the knowledge field that has received much attention in manufacturing. Chow et al. [23] define knowledge management as activities that enable the creation, storage, distribution, and application of knowledge. Thus, knowledge management relates to business processes where the creation, transfer, and adaptation of knowledge are basic elements [24].

16.5 Basic Dimensions of Knowledge

Weinert [25] distinguishes "general world" knowledge from "arbitrary, specialized" knowledge as follows. General world knowledge, which overlaps considerably with what is defined as "crystallized intelligence," is generally measured by vocabulary tests that are part of many intelligence measurements. Arbitrary, specialized knowledge is knowledge necessary for meeting content-specific demands and for solving content-specific tasks. The arbitrary, specialized knowledge is typically knowledge you need in non-predictable situations where actions or patterns are difficult to program. This could for example be where manufacturing is closely built on customers changing requirements, where the quality of input in a process vary considerably, and where diffuse social factors and human interaction are premise providers for products and processes.

Arbitrary knowledge is very important in emergency and patient treatment in hospitals, but also in manufacturing especially in craft manufacturing. Arbitrary knowledge could also be important in innovation and creativity processes. In these situations intuition and improvisation become keys for success and experience an important knowledge skill. This knowledge is often difficult to communicate outside their context in a written form. However, you often see that operators and craftsmen use some kind of notebooks to capture the particularities of the situations and to develop their arbitrary knowledge.

Arbitrary knowledge is essential for most kind of organizations, but is also difficult to capture and distribute. Many attempts have been made to enable knowledge sharing of this kind and making it more structured and mathematical. Artificial intelligence and semantic web are key examples of these attempts. However, removing the contextual and personal aspects of arbitrary knowledge is still difficult, probably impossible and sharing this type of knowledge is a challenge between organizational units within a company and certainly between companies and their outsourced suppliers.

16.6 Tacit and Explicit Knowledge

A knowledge dimension that is similar to the 'general world' versus 'arbitrary specialized', and that has been emphasized in the literature is 'tacit knowledge' versus 'explicit knowledge' [9]. Nonaka [26] describes this as an epistemological dimension of organizational knowledge creation. This dimension suggests that there are two types of knowledge: tacit, which is embedded in the human brain and cannot be expressed easily, and explicit knowledge, which can be easily codified. Both types of knowledge are important, but Western organizations have focused largely on managing explicit knowledge.

Explicit knowledge, also known as codified knowledge, corresponds to the information and skills that are easily communicated and documented, such as processes, templates, and data that are captured in media. Explicit knowledge, thus, is easier to reuse across an organization. Explicit knowledge or codified knowledge is transmittable in formal systematic language. In companies with close relationships to other companies, a third type of knowledge is also important, namely 'cultural knowledge' [27].

Tacit knowledge is highly personal knowledge that is gained through experience and largely influenced by beliefs, perspectives, and values embedded in the individual experiences of workers. It is deeply rooted in action, commitment, and involvement within a specific context. It has a personal quality which makes it hard to communicate and explain. Tacit knowledge involves both cognitive and technical elements [26]. The cognitive element is related to how working models, including schemata, paradigms, and beliefs, among others, help persons to perceive and define 'their world'. The technical element covers concrete know-how, crafts, skills, and techniques that apply to a concrete context. Arbitrary knowledge is often tacit since it is context specific and often not easy to communicate. However, arbitrary knowledge can also be explicit and codified.

The notion of 'embeddedness' was introduced into the knowledge field by Granovetter [28]. Focusing on the social and institutional arrangements, embedded knowledge is reflected in relationships between technologies, roles, procedures, routines, and so on [29].

Even if knowledge is to a large extent tacit in an organization, it can often also be made more explicit if the contexts are understood by others and it is possible to structure and codify it, if not in absolute and mathematical terms, then perhaps in written text or through indicators. From this we understand that it is difficult to make a clear distinction between some forms of tacit and explicit knowledge. This is also a major reason why companies in their continuous improvement and even R&D activities often put much effort into structuring and codifying tacit knowledge. This also includes defining arenas for knowledge sharing such as 'stand up meetings', 'monday planning meetings' and so on at different levels in the organizations. It can even extend to creating a common format, language or semantics for articulation and communication of key concepts. In the business process improvement and quality improvements toolbox there are many examples of templates and methodologies for structuring tacit knowledge. These methods range from using 'post its' to more advanced lean methodologies such as the 'A3 process'.

Much knowledge can be recorded, but, nevertheless, the assets of an organization are increasingly in its employees and their tacit knowledge. "It is more important for organizations to exploit and manage their intangible assets in contrast to their physical assets" [30]. Management of intangible assets includes knowing who knows what and is an important part of management. One study found, for example, that "system engineers need to apply just as much effort and attention to determine who to contact in the organization in order to get the job done" [31]. Another study found that people in systems engineering organizations spent about 40% of their time in searching for and accessing different types of information related to their projects [32].

16.7 Knowledge Bases and Innovation

The ontological dimension of knowledge can be understood as the process that organizationally amplifies and transfers knowledge [33]. This process will often be related to innovation processes and improvements across units and organizations. This could involve suppliers or customers in a value chain. It could also involve other more 'distant' stakeholders such as universities and R&D partners, governmental bodies, and so on. This process might be formal through a structured project model such as the ones presented in Chap. 15. In day-to-day operations, it might occur in informal communities of interaction or in informal contact between people.

A critical factor is an organizations' 'absorptive capacity' [34]. An organization's absorptive capacity depends on individual capacities, on transfers of knowledge across environmental boundaries and across sub-units. Zahra and George [35] specify four facets of absorptive capacity: acquisition, assimilation, transformation and exploitation. When absorption limits exist, they provide one explanation for firms to develop internal R&D capacities. Perhaps even more important for increasing absorptive capacity is a culture and strategy for lifelong learning, where individuals are trained to continuously improve or challenge knowledge, competencies and skills. In this way individual capacities are increased. Knowledge brokers and R&D departments can not only conduct development along lines that they are already familiar with, but they have formal training and external professional connections that make it possible for them to evaluate and incorporate externally generated technical knowledge into the firm. They might also help and guide employees in finding and sharing relevant knowledge. In other words, a partial explanation for R&D investments by firms is to work around the absorptive capacity constraint.

As pointed out by Asheim and Gertler [36], there are always both tacit and codified knowledge involved and needed in the process of knowledge creation and innovation. This knowledge represents a companies, value chain, or entire industries knowledge base. Different industries and technological sectors may follow different paths of innovation [37]. These heterogeneous trajectories are to a large extent historically determined by the differences in the underlying knowledge base. Asheim and Gertler [36] have developed frameworks often applied for analysis of industrial clusters. They describe how the innovation process of industries and companies could be regarded as strongly shaped by their specific knowledge base. The two types of knowledge base vary systematically between industries, but also depend on company strategies:

- *Synthetic knowledge base*—prevails in industrial settings where innovation takes place mainly through the application of novel combinations of existing knowledge. This is often related to incremental product/process development related to the solution of specific problems, and is normally tacit, based on concrete know-how and practical skills
- *Analytical knowledge base*—dominates where scientific knowledge is highly important. It is often based on formal models, explicit knowledge and rational processes. Such knowledge is likely to lead to radical innovations in the form of new products or processes.

A synthetic knowledge base refers to industries and companies where innovation takes place mainly through the application of knowledge or through recombination of existing knowledge in new ways [37]. Examples of fields that rely mainly on a synthetic knowledge base include plant engineering, food processing, telecommunications, and advanced industrial machinery and production systems. The synthetic knowledge base is particularly important in lean manufacturing and craft manufacturing. These are industries characterized by a more incremental way of innovation, dominated by the modification of existing products and processes. The intellectual challenge is geared towards constructing and running complex functional systems shaped as producible and useful artifacts [38].

Normally, all industries and companies have an analytical as well as a synthetic knowledge base, and use tacit as well as explicit knowledge. What is significant is how they are balanced and applied, for example, according to strategies. Examples of technological sectors that rely mainly on an analytical knowledge base include biotechnology, nanotechnology, information technology and genetics. Both basic and applied research becomes relevant activities, and the intellectual challenge is to understanding natural systems by discovery and application of natural ways for a systematic development of products and processes.

Figure 16.1 illustrates how innovations are normally based on a combination of tacit—and explicit knowledge from the synthetic- and analytical knowledge base. This combination of knowledge will be a necessity in innovations since the explicit knowledge need to be 'translated' into a particular context using tacit knowledge. The tacit knowledge on the other hand will often be more useful when it is supported by explicit facts and details.

The increased focus on human aspects of innovation indicates that the contextual, tacit knowledge, and the synthetic knowledge base, also increase in importance in a learning economy. This aspect is reflected in the IMS 2020 'Roadmap on

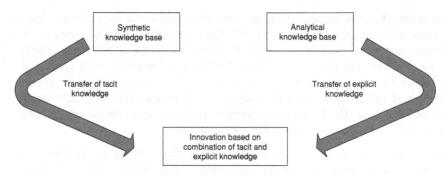

Fig. 16.1 Innovations through tacit and explicit knowledge

Innovation, Competence Development and Education', where human resources for science and technology are described as vital to innovation and economic growth because highly skilled people create and diffuse innovations [39]. In most countries, the demand for skilled workers is expected to increase owing to real growth in R&D and the growing application of advanced technologies in many industries:

> Manufacturing has moved from a pure technology view to a view integrating technology, business and management. The extended view on manufacturing reflects a need for a new competence for the industry. The future manufacturing engineer must be trained for the integrated view and must at the same time manage the societal needs for sustainability and environmental protection [40]

Nonaka [26] has stated that to facilitate true learning organizations, tacit knowledge must be the cornerstone of future investigation. According to him tacit knowledge becomes an important part in innovation and strategic development especially when it is related to more formal structures and is codified. Von Krogh et al. [41] provide five knowledge enablers as a means to develop the power of tacit knowledge, and that it is through these knowledge enablers that sharing of knowledge can occur and that true organizational improvement will happen:

1. Install a knowledge vision
2. Manage conversations
3. Mobilize knowledge activists
4. Create the right context
5. Globalize local knowledge.

However, it is not evident how to apply these knowledge enablers as they appear more as principles than methods or tools. This is one of the reasons for the previously described development (or attempts to develop) of tools, ICT applications and methodologies for knowledge sharing. However, it is an illusion to think that all tacit knowledge could be codified and made explicit to fit a structured innovation process. Tacit knowledge also represents an enabler for improved innovation processes through richer discussion and better analysis through a contextual understanding of knowledge among knowledge workers.

A new concept to be introduced by Chesbrough [42] is that of "open innovation". The central idea of open innovation is that in a world consisting of widely distributed knowledge, organizations like enterprises cannot afford to rely entirely on their own knowledge creation process. Instead they should buy or license processes or inventions (e.g. patents) from other companies. In addition, internal innovations which are not used in business could be taken on outside the enterprise (e.g. by licensing or joint ventures). Open innovations has become popular within software development, where the operating system LINUX is a well-known example. Open innovation has also been used to acquire feedback and input from markets and customers for new products and concepts.

In recent years a new field has emerged that, blends the skills of computer science, statistics, artistic design and storytelling, and aimed at displaying and making massive amounts of data easily accessible [43]. Yau [44] describes this as a new medium that presents "meaty information in a compelling narrative: something in-between the textbook and the novel" [44, p. 13]. Displaying information can enable people to understand complex matters and find creative solutions. These ways of displaying information could enable the transfer of explicit knowledge, but also making tacit knowledge more understood in other contexts. These approaches have the early stages of innovation in focus. The early stage of innovation is characterized by high uncertainties and the constant generation of new and relevant knowledge [45]. The knowledge is generated and immediately used in non-linear work steps. The complexity of these work processes cannot be described appropriately in a quantifiable model [46].

References

1. Rasmussen B (2004) Organizing knowledge work(ers): the production of commitment in flexible organizations. In: Carlsen A, Klev R, von Krogh G (eds) Living knowledge. The dynamics of professional service work. Basingstoke, Palgrave Macmillan, pp 67–87
2. Livingston P (2003) Thinking and being: Heidegger and Wittgenstein on machination and lived-experience. Inquiry 46(3):324–345
3. Dreyfus H, Dreyfus S (1991) Intuitive ekspertise: Den bristende drøm om tenkende maskiner. DTB, Copenhagen (trad. "Intuitive expertice: the ruptured dream of the thinking machine")
4. Lin C, Hung HC, Wu JY, Lin B (2002) A knowledge management architecture in collaborative supply chain. J Comput Inf Syst 42(5):83–94
5. Wilson JR (2003) Support of opportunities for shopfloor involvement through information and communication technologies. AI Soc 17(2):114–133
6. Filos E, Banahan E (2001) Towards the smart organization: an emerging organizational paradigm and the contribution of the European RTD programs. J Intell Manuf 12(2):101–119
7. Myers D (2004) Psychology, 7th edn. Hope College, Michigan
8. Livingston P (2003) Thinking and being: Heidegger and Wittgenstein on machination and lived-experience. Inquiry 46(3):324–345
9. Polyani M (1966) The tacit dimension. Routledge & Kegan Paul, London

10. Nonaka I, Takeuchi H (1995) The knowledge creating company. Oxford University Press, New York
11. Lai H, Chu TH (2002) Knowledge management: a review of industrial cases. J Comput Inf Syst 42(5):26–39
12. Sveiby KE (1997) The new organizational wealth: managing & measuring the knowledge-based asset. Berret-Koehler Publishers, San Francisco
13. Dretske F (1981) Knowledge and the flow of Information. MIT Press, Cambridge
14. Winterton J, Le Deist FD, Stringfellow E (2005) Typology of knowledge, skills and competences: clarification of the concept and prototype. Centre for European Research on Employment and Human Resources Groupe ESC Toulouse, Research report elaborated on behalf of Cedefop/Thessaloniki, Final draft, (CEDEFOP Project No RP/B/BS/Credit Transfer/005/04)
15. Bloom BS (1976) Human characteristics and school learning. McGraw-Hill, New York
16. Bloom BS, Hastings JT, Madaus GF (1971) Handbook on formative and summative evaluation of student learning. McGraw Hill, New York
17. Bloom BS, Mesia BB, Krathwohl DR (1964) Taxonomy of educational objectives (two vols: The affective domain and the cognitive domain). David McKay, New York
18. Proctor RW, Dutta A (1995) Skill acquisition and human performance. Sage Publications, London
19. Collin A (1989) Manager's competence: rhetoric, reality and research. Personnel Rev 18(6):20–25
20. Crawford L (2000) Project management competence for the new millennium. In: Proceedings of 15th world congress on project management, IPMA, London, England, May 2000
21. Punnitamai W (2002) The application of competency modeling for human resource management: a holistic inquiry. Thai Journalof Public Administration Sept–Dec:113–132
22. PMI (2002) Project manager competency development framework. Project Management Institute, Pennsylvania
23. Chow HKH, Choy KL, Lee WB, Chan FTS (2005) Design of a knowledge-based logistics strategy system. Expert Syst Appl 29:272–290
24. Gunasekaran A, Ngai EWT (2007) Knowledge management in 21st century manufacturing. Int J Prod Res 45(11):2391–2418
25. Weinert FE (1999) Concepts of Competence. Manx Planck institute for psychological research, Munich [Published as a contribution to the OECD project. Definition and selection of competencies: Theoretical and conceptual foundations (DeSeCo). Neuchâtel: DeSeCo.]
26. Nonaka I (2004) A dynamic theory of organizational knowledge creation. In: Starkey K, Tempest S, McKinlay A (eds) How organizations learn. Managing the search for knowledge. Thomson Learning, London, pp 165–201
27. Bontis N, Chun WC (2002) The strategic management of intellectual capital and organizational knowledge. Oxford University Press, New York
28. Granovetter M (1985) Economic action and social structure: the problem of embeddedness. Am J Sociol 91(3):481–510
29. Bandaracco JL (1991) The knowledge link: how firms compete through strategic alliances. Harvard Business School Press, Boston
30. Tiwana A (2000) The knowledge management toolkit: practical techniques for building a knowledge management system. Prentice Hall, New Jersey
31. Perry DE, Staudenmayer NA, Votta LG (1994) People, Organizations, and Process Improvement. IEEE Softw 11(4):36–45
32. Henninger S (1997) Case-based knowledge management tools for software development. J Autom Soft Eng 4(3):319–340
33. Wilber K (2000) Integral psychology: consciousness, spirit, psychology, therapy. Shambhala Publications, Boston
34. Cohen WM, Levinthal DA (1990) Absorptive capacity: a new perspective on learning and innovation. Adm Sci Q 35(1):128–152

35. Zahra SA, George G (2002) Absorptive capacity: a review, reconceptualization, and extension. Acad Manage Rev 27(2):185–203
36. Asheim BT, Gertler M (2005) The geography of innovation—regional innovation systems. In: Fagerberg J, Mowery D, Nelson R (eds) The oxford handbook of innovation. Oxford university Press, Oxford, pp 291–317
37. Pavitt K (1984) Sectoral patterns of technical change: towards a taxonomy and a theory. Res Policy 13:343–373
38. Asheim BT, Coenen L (2005) Knowledge bases and regional innovation systems: Comparing Nordic clusters. Res Policy 34(8):1173–1190
39. Rolstadas A Moseng B, Vigtil A, Osteras T, Fradinho M, Carpanzano E, Bromdi C, (2010) Action Roadmap on Key Area 5. IMS2020 project report. Trondheim, Norway
40. IMS2020 (2010) Roadmap on innovation, Competence Development and Education. Available on: http://www.ims2020.net/, Accessed Oct 2012, pp 26–27
41. Von Krogh G, Ichijo K, Nonaka I (2000) Enabling knowledge creation. How to unlock the mystery of tacit knowledge and release the power of innovation. Oxford University Press, Oxford
42. Chesbrough HW (2003) The era of open innovation. MIT Sloan Manage Rev 44(3):35–41
43. Tufte E (2001) The visual display of quantitative information. Graphics Press, Cheshire
44. Yau N (2009) Seeing your life in data. In: Segaran T, Hammerbacher J (eds) Beautiful data: the stories behind elegant data solutions. O'Reilly Media Inc., Canada, pp 1–15
45. The Economist (2010a) Two and a half cheers for sticks and carrots. 16th Jan, p 62
46. Akin Ö (1979) Exploration of the design process. Des Methods Theor 13(3/4):115–119

Chapter 17
Knowledge Creation

Abstract Knowledge production and capture includes all of the processes involved in the acquisition and development of knowledge. Knowledge integration/codification involves the conversion of knowledge into accessible and applicable formats. Knowledge transfer/use includes the movement of knowledge from its point of generation or codified form to the point of use. One of the reasons that knowledge is such a difficult concept is because this process is systemic and often discontinuous. Many cycles are occurring concurrently in businesses. These cycles feed on each other. Knowledge interacts with information to increase the space of possibilities and provide new information, which can then facilitate generation of even more new knowledge. The ability to share knowledge across national borders is an important reason behind the formation of multinational corporations. Improving knowledge creation is to a large extent a question of enabling these processes and removing barriers for sharing and transferring knowledge. Nonaka and Takeuchi have developed a model for organizational knowledge creation that emphasizes tacit and explicit knowledge in different modes. Knowledge transfer is the basis for most development projects and subsequent mobilization of tacit knowledge.

17.1 Putting Bits and Pieces Together

Knowledge has to come from somewhere, or could it just appear from someone or something that 'suddenly sees the light'? This could easily turn out to be a question of existential or religious nature and we will not dig too much into that. In this book we will see knowledge creation as the process of coupling and sharing knowledge into a bigger whole. Thus the definition by Nonaka and Krogh [1] of organizational knowledge creation appeals to us:

A. Rolstadås et al., *Manufacturing Outsourcing*,   157
DOI: 10.1007/978-1-4471-2954-7_17, © Springer-Verlag London 2012

Organizational knowledge creation is the process of making available and amplifying knowledge created by individuals as well as crystallizing and connecting it to an organization's knowledge system [1, p. 635].

Knowledge production and capture includes all the processes involved in the acquisition and development of knowledge. Knowledge integration/codification involves the conversion of knowledge into accessible and applicable formats. Knowledge transfer/use includes the movement of knowledge from its point of generation or codified form to the point of use. One of the reasons that knowledge is such a difficult concept is because this process is systemic and often discontinuous. Many cycles are concurrently occurring in businesses. These cycles feed on each other. Knowledge interacts with information to increase the space of possibilities and provide new information, which can then facilitate generation of new knowledge.

The ability to share knowledge across national borders is the prime reason behind the formation of multinational corporations (MNCs) [2]. Within the MNC network, knowledge may be created in one location, and put to productive use in many other locations [3, 4]. The key challenges are in the willingness of a unit to share knowledge with other units [5, 6], the recipient's ability or willingness to absorb [5, 7, 8] and in the relationship between the sender and receiver.

17.2 The Trade-Off Barrier

Improving knowledge creation is to a large extent a question of identifying and removing barriers for sharing and transferring knowledge. Earlier research uncovered a number of factors that impact on knowledge transfer between units in global companies and that often represent barriers to knowledge transfer [9]. One barrier can be "a probable trade-off between resources deployed in knowledge development and resources deployed in transfer to other units" [10, 11]. In a distributed manufacturing strategy this could mean that a unit can emphasize the use of limited resources in local R&D-activities and projects aiming to develop capabilities and specialized knowledge in for example a outsourced manufacturing unit. This local knowledge creation could then be preferable to an often difficult and resource demanding knowledge transfer where the outcome for the local unit could be diffuse or even negative as they could lose their exclusive knowledge base.

Nonaka and Takeuchi [12], in their model for organizational knowledge creation, emphasize person-oriented and informal mechanisms as absolutely necessary to shift tacit knowledge from one organizational boundary to another. According to Reger and Gerybadze [13], this means that the organizational knowledge creation process can only be successful if tacit knowledge is transferred and coordinated using appropriate instruments. However, internal market mechanisms are useful only with regard to the exchange of explicit knowledge; vice versa, if they are applied alone, explicit and not tacit knowledge is transferred.

17.3 Conflicts of Interests

Companies with a strategy for distributed manufacturing where central control and coordination is low have many incentives for local initiatives that can enhance knowledge creation. However, "conflicts of interest are likely to emerge between units" [6, 14, 8] especially from outsourced units. Knowledge can represent unique capabilities and an asset that companies lose if they give it away to other companies. The unique knowledge can represent one of the main assets for a unit or subcontractor in a battle for power in the supply chain, or for having the possibility to attain other customers. Even if there are obvious potential barriers for knowledge transfer, there will also be knowledge spillovers within the organization to the other actors in the supply chain.

Based on agency theory and socialization theory Björkman et al. [8] explored the impact of organizational mechanisms on inter-unit knowledge flows in 134 Finnish and Chinese MNC's. Their findings indicated that MNCs can influence inter-unit knowledge transfer by specifying the objectives of the subsidiary and by utilizing corporate socialization mechanisms. They found no impact of management compensation systems and the use of expatriate managers on knowledge transfers from foreign subsidiaries to other parts of the MNC.

17.4 Knowledge Conversion

One barrier for knowledge transfer that has been previously discussed is "the character of the knowledge itself, its senders and recipients, and the relationship between them" [12, 15–17]. There are many challenges related to explicit knowledge when it comes to language, systems, coding/decoding capacities, and so on. These challenges are obvious between companies, but improved systems and enabling technologies have reduced these barriers. This has improved collaboration and streamlining within supply chains. New technologies have also enabled more flexible supply chain networks, which has been manifested through an adaptive manufacturing paradigm. A field where new technologies have been important has been within sustainability and environmental impact of manufacturing. Through open sources or other databases and systems the environmental footprint is now much easier to identify and describe.

Most attention has recently been on how to transfer the context specific tacit knowledge. Due to the problem of separating knowledge from the unit that carries the knowledge and adding it to another actor's knowledge base, it has been argued that idiosyncratic, specific, tacit, and/or non-codified knowledge is difficult to transfer from one unit to another [5, 10, 17–19]. The challenges will normally increase where units, or collaborating companies, develop different business cultures, work according to different manufacturing principles (paradigms), and there is spatial distance.

Different contexts for senders and receivers represent the main barrier to knowledge transfer in addition to the difficulties to articulate and present the tacit knowledge. The knowledge creation spiral of Nonaka and Takeuchi [12] presented and initially discussed in Fig. 5.1 has been subject to much of the theory development within knowledge creation, and the focus is mainly on the tacit knowledge. Thompson [20] suggested that organizational hierarchies tend to cluster most closely in groups with reciprocal interdependencies compared to those with sequential interdependencies or pooled interdependencies. It follows that focusing on knowledge as context-specific, but also relational [21], at least three mechanisms for the coordination of knowledge can be hypothesized:

1. *Changing the context for tacit knowledge, conversion to a broader context, which also means a broader synthetic knowledge base.* This conversion [22] is normally difficult as it requires changing values, norms and perspectives. Structural decisions could influence the context, as could, for example, the management principles and style. The literature identifies a particularly need for 'knowledge pipelines' not only bringing in new explicit knowledge, but also new perspectives [23]. Job rotation, exchange programs, project orientation, and flexible organizational structures are example of relevant means. A PMO (Project Management Office) could also take on this role as knowledge pipeline provider
2. *Making tacit knowledge explicit, through different ways of structuring and codifying knowledge or decoding, making it comprehensible and relevant outside the original context.* Storytelling is a means, but one that is not frequently used [24], and a more typical approach is through codification
3. *Making explicit knowledge tacit, knowledge created in another context, but which is basically explicit, for example R&D knowledge, should be considered relevant in an operational context.* This means that knowledge has to be transferred in a way that reflects the context in which it is supposed to be adapted. In this situation, storytelling has also been referred to as a potential means. Another approach is through job rotation, for example where knowledge is transferred through persons who practice and use it in the new context. A defined role as a knowledge broker, a catalyst for new knowledge, has also been referred to as a means for knowledge transfer [25].

17.5 How to Integrate Knowledge Creation into the Development Model

Since knowledge creation and knowledge transfer is the basis for most development projects and innovations we need project models or product development models that mobilize tacit knowledge and open it up for learning. It's not yet clear how to develop and apply such a model, but at least it would be important to avoid product development models that inherently exclude tacit knowledge.

Fuzzy logic [26] has been invented specifically for describing 'fuzziness' of systems and by association knowledge in precise mathematical terms. In the knowledge creation process, we are faced with a necessity to combine sophisticated mathematical knowledge (explicit) about the analyzed systems with informal expert knowledge (tacit knowledge). One big problem with interpreting commonsense knowledge in precise mathematical terms is that the words that experts use to describe their knowledge are often not precise, they are 'fuzzy',—approximate rather than fixed and exact. Others have developed methodologies to reformulate the abstract mathematical knowledge in more understandable intuitive terms. In this sense for a metric or indicator to be 'natural' means that the metric can be interpreted in commonsense terms. Fuzzy logic methodologies often mean computational methods tolerant to sub-optimality and impreciseness (vagueness) and giving quick, simple and 'sufficiently good solutions' [27].

BEEM (Business Effect Evaluation Methodology) [28] is a way to capture strategic aspects of for example a product or process development project. The methodology aims through workshops and discussions to structure and code knowledge that was initially tacit. BEEM captures in a structured way additional knowledge to the more explicit cost benefit evaluation criteria often used in project decision gates. The methodology is described in more detailed in Sect. 19.3.

The most problematic challenge to knowledge management is that most of the knowledge in manufacturing is tacit and will never become explicit. It will remain tacit because there is no time to make it explicit. There are very few approaches and tools for turning tacit knowledge into explicit knowledge, and most of the tacit knowledge is tacit in the most extreme way. Therefore, it is difficult to express it and make it explicit. A way to address this problem can be to develop a knowledge sharing culture, as well as technology support for knowledge management, never forgetting that the main asset of the organization is its employees.

Even more difficult than just codifying the tacit knowledge could be to establish the mechanisms and process of creativity and learning. Ideation is a concept for a creativity process used by companies such as Starbucks and Best Buy. Ideation is basically a circular process that differs significantly from the sequential process approach that most people are trained in [29]. Ideation is an inter-disciplinary and cross-organizational process that requires a certain degree of a common neutral language. This creativity process includes, generating, developing, and communicating new ideas, where an idea is understood as a basic element of thought that can be either visual, concrete, or abstract [30]. Ideation is all stages of a thought cycle, from innovation, to development, to actualization [31].

The circular ideation process is illustrated in Fig. 17.1 where we see how ideas are developed from an information search, evaluated, form the basis for teambuilding, idea development and so on. This circular process has much in common with Deming's PDCA cycle. Ideation is a method that aims to capture both tacit and explicit knowledge through information search but also through team discussions and evaluation. The circular ideation process builds teams and creates new knowledge in a way that could enable knowledge transfer especially of the tacit kind. Figure 17.2 shows how ideation is a part of knowledge creation and knowledge transfer in the

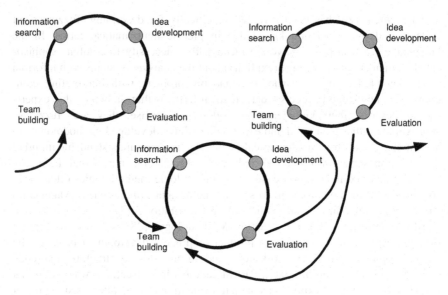

Fig. 17.1 The ideation process (IMS 2020) [32]

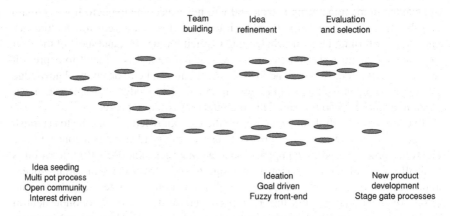

Fig. 17.2 New product development through ideation (IMS 2020) [32]

whole innovation process, from idea seeding and generation to product and process development which then leads to operations, marketing and sales.

Figure 17.3 is in fact the same product development model as the one presented in Fig. 15.5 for the supplier in the automotive industry. But whereas the illustration in Fig. 15.5 focuses on decisions and how the different alternatives are evaluated, Fig. 17.3 is more focused on the knowledge creation and organizational learning aspects that are the basis for the different concepts being developed and evaluated. The circular learning processes have much in common with the ideation process. We see that the learning process progresses to the 'Design Freeze' in decision gate

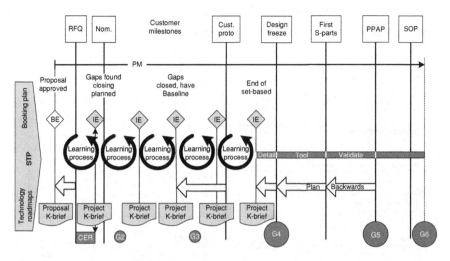

Fig. 17.3 Repeated learning cycles in a product development model

G4. However, there will typically be a continuous learning process thereafter this decision gate covering elements other than product design. For example it could cover process design, operational decisions and learning related to organizational development, training, sales and distribution.

In the past, cross-border R&D was largely aimed at adapting products and services to the needs of host countries. It was carried out close to 'lead users' in order to adapt products and processes to local conditions. The basis for R&D activities were normally done at the mother plant that also supported the local manufacturing operations of multinational enterprises (MNEs). At present, MNEs seek not only to exploit knowledge generated at home and in other countries, but also to source technology internationally and tap into centers of increasingly multidisciplinary knowledge worldwide.

Globalization of R&D and manufacturing education seems to offer career paths with better job offers for students and more opportunities for research projects. It also create and a more realistic understanding of the problems of manufacturing in a global environment [33]. This globalization of career paths and exchanges of people could improve the learning processes in Fig. 17.3, but the ideation process illustrated in Fig. 17.1 is obviously difficult even if there is an increase in the mobility of people within the company or along a global supply chain. The ideation process requires some kind of spontaneous team building and exchange of tacit knowledge that is still difficult in the world of virtual societies.

In tightly connected supply chains as we see in the automotive industry there will often develop relations between key personnel and resources on the interfaces of organizational boundaries. This could for example be the case between the OEM and the first tier suppliers in the early concept phases of product development. Integrated project development models such as the one in Figs. 15.5 and 17.3 could enable an ad hoc ideation process and set direction for the innovation process.

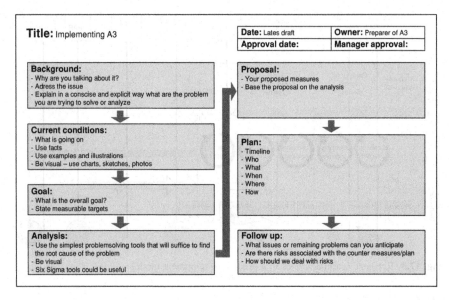

Fig. 17.4 Example of an A3 template

17.6 Process Improvement and Knowledge

It is difficult to make a clear cut distinction between the terms, research, development, process improvement, quality improvement, and so on. To which extent we are talking about changes of substantial character can help us to attempt to separate the terms. Simple process improvement in one industry could be research in another. For example the use of robots and automated processes in the leisure boat industry would be regarded as radical R&D projects. For a supplier in the automotive industry where processes are often standardized, this would often be simply a question of application adjustment and implementation. Even if the creativity process has to come up with completely new ideas and knowledge that might be less resource demanding, we still normally need to deal with a combination of tacit- and explicit knowledge.

Many of the quality improvement methods and tools aim to engage people, mobilize knowledge and find ways to share knowledge. Today this often means to make tacit knowledge from the shop floor a key element in quality improvement. This has been emphasized for example in lean manufacturing and many tools have been developed to enable this process.

One such method is the A3 process, which has proved to be a powerful method that systematically guides problem-solvers through a rigorous process, documents the key outcomes, and suggests improvements [34]. It is a tool used by Toyota to implement the PDCA process [35] and is described by a simple storyboard that tells the whole story of an improvement event on one 11×17 cm sheet of paper (see Fig. 17.4). The left-hand side is used to define the problem, while the right-hand side is used to suggest solutions. The A3 report helps to identify the root causes, leading to in-depth solutions, and bundling them up in a simple and easy to share format.

References

1. Nonaka I, von Krogh G (2009) Tacit knowledge and knowledge conversion: controversy and advancement in organizational knowledge creation theory. Organ Sci 20(3):635–652
2. Gupta AK, Govindarajan V (2000) Knowledge flows within multinational corporations. Strateg Manag J 21(4):473–496
3. McCann P, Mudambi R (2005) Analytical differences in the economics of geography: the case of the multinational firm. Environ Plann A 37(10):1857–1876
4. Mudambi R, Navarra P (2004) Is knowledge power? Knowledge flows, subsidiary power and rent-seeking within MNEs. J Int Bus Stud 35(5):385–406
5. Szulanski G (1996) Exploring internal stickiness: impediments to the transfer of best practice within the firm. Strateg Manage J 17(Winter Special Issue):27–43
6. Forsgren M (1997) The advantage paradox of the multi-national corporation. In: Bjorkman I, Forsgren M (eds) The nature of the international firm: nordic contributions to international business research. Copenhagen DK, Copenhagen Business School Press, pp 69–85
7. Allen T (1977) Managing the flow of technology: technology transfer and the dissemination of technological information within the R&D organization. MIT Press, Cambridge
8. Cohen WM, Levinthal DA (1990) Absorptive capacity: a new perspective on learning and innovation. Adm Sci Q 35(1):128–152
9. Bjorkman I, Barner-Rasmussen W, Li L (2004) Managing knowledge transfer in MNCS: the impact of headquarters control mechanisms. J Int Bus Stud 35:443–455
10. Forsgren M, Johansson J, Sharma D (2000) Development of MNC centres of excellence. In: Holm U, Pedersen T (eds) The emergence and impact of MNC centres of excellence. Macmillan, London, pp 45–67
11. Szulanski G (2003) Sticky knowledge: barriers to knowing in the firm. Sage, London
12. Nonaka I, Takeuchi H (1995) The knowledge creating company. Oxford University Press, New York
13. Reger G, Gerybadze A (1997) New coordination mechanisms and flexible lateral organization within transnational corporations—Discussion paper on international management and innovation, Discussion-paper 97-04, Stuttgart: Hohenheim University. Available at https://www.uni-hohenheim.de/ innovation/downloads-frei/discussion_papers/ DIMI974.pdf. Accessed Nov 2009
14. Birkinshaw J, Hood N (1998) Multinational subsidiary evolution: capability and charter change in foreign-owned subsidiary companies. Acad Manag Rev 23(4):773–795
15. Levitt B, March JG (1988) Organizational learning. Ann Rev Soc 14:319–340
16. Zander U, Kogut B (1995) Knowledge and the speed of the transfer and imitation of organizational capabilities: an empirical test. Organ Sci 6(1):76–92
17. Hansen MT (1999) The search-transfer problem: the role of weak ties in sharing knowledge across organization subunits. Adm Sci Q 4(1):82–111
18. Grant RM (1996) Prospering in dynamically-competitive environments: organizational capability as knowledge integration. Organ Sci 7(4):375–387
19. Spender JC (1996) Making knowledge the basis of a dynamic theory of the firm. Strateg Manage J 17(Winter Special Issue):45–62
20. Thompson JD (1967) Organizations in actions: social science bases of administrative theory. McGraw-Hill Book Company, New York

21. Berger P, Luckman T (1967) The social construction of reality. A treatise in the sociology of knowledge. Penguin Press, Harmondsworth
22. Nonaka I, Konno N, Toyama R (2001) Emergence of "Ba". A conceptual framework for the continuous and self-transcending process of knowledge creation. In: Nonaka I, Nishiguchi T (eds) Knowledge emergence: social, technical, and evolutionary dimensions of knowledge. Oxford University Press, New York
23. Maskell P, Bathelt H, Malmberg A (2006) Building global knowledge pipelines: the role of temporary clusters. Eur Plann Stud 14(8):997–1013
24. Røyrvik EA, Bygdås AL (2004) Knowledge hyperstories and context-sensitive knowledge enabling. In: Carlsen A, Klevand R, von Krogh G (eds) Living knowledge. The dynamics of professional service work. Macmillan, Palgrave, pp 184–203 Roth 2007
25. Zadeh LA (1965) Fuzzy sets. Inf Control 8(3):338–353
26. Novák V (1989) Fuzzy sets and their applications. Adam Hilger, Bristol
27. Henriksen B, Røstad CC (2010) Evaluating and prioritizing projects—Setting targets. The business effect evaluation methodology BEEM. Int J Manag Projects Bus 3(2):275–291
28. Leifer R, McDermott CM, O'Connor GC, Peters LS, Rice M, Veryzer RW (2000) Radical innovation: how mature companies can outsmart upstarts. Harward Business School Press, Boston
29. Johnson B (2005) Design ideation: the conceptual sketch in the digital age. Des Stud 26(6):613–624
30. Graham D, Bachmann T (2004) Ideation: the birth and death of ideas. Wiley, NY
31. IMS (2010) 2020 Roadmap on Innovation, Competence Development and Education 15 July 2010
32. Stephan KD, Vedaraman S (2005) Globalizing manufacturing engineering education. Technol Soc Mag 24(3):16–22
33. Sobek DK (1997) Principles that shape product development systems: A Toyota-Chrysler comparison. PhD Thesis, The University of Michigan, Ann Arbor
34. Kennedy MN (2010) Knowledge based product development—Understanding the true meaning of lean in product development. Presentation at the seminar "Knowledge based development forum 27th–28th Jan, Kongsberg, Norway

Chapter 18
Knowledge Transfer and Distance

Abstract One of the basic considerations for companies when setting up global units, outsourcing units, or more generally entering into global markets, is what distance may mean when dealing with knowledge. There is no doubt that ICT has changed or reduced these challenges and in turn increased globalization in manufacturing. However, distance is not only a question of space and geography, it is also a question of cultural, political, and infrastructural differences and obstacles. These other facets of 'cognitive' distance will tend to have some kind of correlation with spatial distance. There is no doubt that new technologies have enabled transfer of explicit knowledge across organizational and geographic boundaries. When we see that distributed innovation processes, bottom up approaches and employees involvement are increasingly important, we also understand that there are spatial challenges related to tacit knowledge. In fact these challenges may have increased.

18.1 What is Distance

One of the basic considerations for companies when setting up global units, outsourcing units, or simply entering into global markets is what distance means when dealing with knowledge. There is no doubt that ICT has changed or reduced these challenges, while consequently increasing globalization choices for manufacturing. Thompson and Fox-Kean [1] have argued that only national boundaries restrict knowledge flows and that there is no strong evidence to support significant sub-national barriers to knowledge transfer. Literature research finds most proponents for knowledge as 'localized' [2, 3]. Hence, distance has normally been seen as having a major negative impact on knowledge transfer where appropriate strategies and tools have to be applied to deal with this challenge. Distance is not only a question of space and geography, but also a question of cultural, political, and infrastructural differences and obstacles. However, these other facets of distance all tend to have some kind of correlation with spatial distance.

18.2 The Cultural Challenge

Hofstede [4] defines culture as consisting of "shared mental programs that condition individuals' responses to their environment". Culture is an important part of the context wherein knowledge is created or transferred. In Sect. 14.5 we see how Hofstede [5] argues for how perceived cultural compatibility is based on the assumption that normative beliefs about culture cannot be divorced from national culture. He identifies five dimensions of particular importance for explaining differences in business culture: "power distance", "individualism versus collectivism", "masculinity versus feminity", "uncertainty avoidance", and "long-versus short-term orientation".

However, business cultures can be developed independently of national cultures. Business cultures, also according to Hofstedes' dimensions, would also be a result of differences in norms and values, resulting from practices, routines, systems, technologies, individuals, stakeholders, and so on. We also experience that differences in business cultures often develop between units within a company, but certainly between the company and an independent outsourced unit.

Context, partly shaped by culture, represents a filter in the screening process and influences what and how we perceive things. We also tend to have better recall of information and knowledge that is inconsistent with our culturally based expectations but will filter out what is incompatible with our views, thus misinterpreting a person's behavior or communication [6]. In the communication process where knowledge is transferred, culture is the structure through which the knowledge is articulated, formulated and interpreted [7]. Thus we need to understand and be knowledgeable about the cultural constructs that affect knowledge transfer.

18.3 The Spatial Challenge of the Portfolio of Innovations

Knowledge and location issues have received much attention within economic geography. Asheim and Gertler [8] argue that the more knowledge intensive an activity becomes, the more geographically clustered it tends to be. According to Porter [9], market, technical, and other forms of specialized knowledge can be assessed better and be achieved at a lower cost when companies are co-located. One example is the knowledge to be gained from 'sophisticated customers' in the cluster.

Based on agency theory and socialization theory Björkman et al. [10] explored the impact of organizational mechanisms on inter-unit knowledge flows in 134 Finnish and Chinese MNCs. They indicated that MNCs can influence inter-unit knowledge transfer by specifying the objectives of the subsidiary and by utilizing corporate socialization mechanisms. They found no impact of management compensation systems and the use of expatriate managers on knowledge transfers from foreign subsidiaries to other parts of the organization.

Due to the problem of separating knowledge from the unit that carries the knowledge and adding it to another actor's knowledge base, it has been argued that idiosyncratic, specific, tacit, and/or non-codified knowledge is difficult to transfer from one unit to another [11–16]. Challenges arise in the willingness of a unit to share knowledge with other units [14, 15], the recipient's ability or willingness to absorb knowledge [14, 17, 18], and in the relationship between the sender and receiver. Nonaka and Takeuchi [19], in their model for organizational knowledge creation introduced in Fig. 5.1 emphasize person-oriented and informal mechanisms as absolutely necessary to shift tacit knowledge from one organizational boundary to another.

In 1999, Gerybadze and Reger [20] made an in-depth analysis of R&D internationalization in 21 large corporations in Europe, Japan and the US. The findings suggested that transnational corporations had tended to consolidate and streamline their organizations after years of distributed R&D activities and globally-dispersed innovation processes. Dispersed innovation processes had resulted in overly complex and unmanageable organizational architecture inducing firms to search for leaner and more effective types of management for their international portfolio of innovation activities. They concluded that the spatial distribution of learning and R&D activities is something different than the spatial distribution of coordination and control. Many companies in their sample had adopted a strategy of multiple centers of learning with one dominant center of coordination, what we have referred to as a mother—satellite structure. It is not evident that these conclusions are still valid and strategies might have changed and new technologies might have increased decentralization or centralization of innovation processes.

We normally have a portfolio of innovation processes ranging from day- to day improvement and quick fixes to extensive large research projects. These innovation processes have different contexts, involving different types of people producing different kinds of knowledge, in short the knowledge creation and transfer processes are different. For example the tacit knowledge could be much more important in continuous improvement than in the large research projects where explicit knowledge are often more important. Consequently we could very well have a situation as described by Gerybadze and Reger where the innovation processes does not necessarily require close collaboration between many people, where explicit knowledge is coordinated and managed centrally. On the other hand in continuous improvement and small scale development projects, the knowledge is more often tacit and the spatial dimension could be much more important. This is illustrated in Fig. 18.1.

What we see is that the difficulties in separating tacit knowledge from the person who carries it gives ride to increasing challenges as the distance increases between the people involved in the knowledge creation. In this sense distance includes both contextual and spatial distance, two aspects that are closely related.

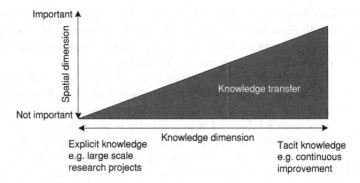

Fig. 18.1 Spatial dimension of knowledge

Where the knowledge is explicit, as we often find in research projects, the spatial distance becomes less important.

18.4 ICT Reduces Distance

ICT has in many ways changed the opportunities and challenges related to distance. There is no doubt that new technologies have enabled transfer of explicit knowledge across organizational and geographic boundaries. Some forms of knowledge can indeed be stored in a pure or media-independent way. We have also seen an increased number of methods and technologies for codifying knowledge into a format that could be more easily transferred through ICT and technologies for rich interactive displays can be generated automatically from stored knowledge. This might take the form of logical propositions, which could be output in graphical, text, or other media to express the same basic knowledge. It might also take the form of tabular data, which can be automatically translated into appropriate graphs or other visualizations [21]. Hypertext has often been considered as a means for sharing knowledge among a community of users.

Many project teams use Web-based platforms to organize their work. Definitions, facts and documents are linked to each other in order to add relevant contextual information to the text itself. This means that there is knowledge that we have before considered as tacit can now be transferred as if it were explicit. However, it still requires considerable time to develop a common understanding during an ad-hoc process and interaction among people is still necessary in most organizations to create this common understanding. Spatial concentration does not fully substantiate the ideas about ICT leading to the dispersal of innovative activity. Rather, this could be explained by the increased importance of tacit knowledge as access to explicit knowledge becomes easier. Since tacit knowledge is heavily imbued with meaning arising from the social and institutional context it is "spatially sticky" [8]. Storper and Venables [22] use the term "buzz" to explain the phenomenon of how this type of localized knowledge circulates through commonly shared frames of experience and understanding. Bathelt et al. [23] argue that geographically clustered companies need access to non-local sources of knowledge. These "pipelines" are essential complements to the local buzz. Without these external knowledge sources there are risks of a "lock-in" situation reducing the ability to improve and adapt to changing conditions.

The rise of ICT networks has provided firms with the logistical potential to better organize and coordinate their R&D activities on a global basis, thus enabling decentralization [24]. However, based on empirical studies, Pauly and Reich [25] argue that the technology generating activities of global companies remain the most domestic of all activities. One explanation could be that the increased ability to codify and transfer knowledge increases the means for the mother company to manage R&D activities. Sawhney and Prandelli [26] describe the potential of ICT to enable knowledge creation and transfer as follows:

> Interactive technologies reduce distances both in time and space, catalyzing knowledge sharing and transfer. Physical distances are often becoming less relevant than cognitive distances. Indeed the digital revolution can be seen as a "cognitive revolution", a revolutionary way to organize and share knowledge... the increasing complexity of the business environment implies a stronger need for knowledge to reduce uncertainty [26, p. 273].

From the above we understand that spatial distance still matters and that this is often due to differences in culture and contextual differences. When we see that in distributed innovation processes, where bottom up approaches and employee involvement are increasingly important, we also understand that there are spatial challenges related to tacit knowledge. In fact these challenges may not have decreased.

References

1. Thompson P, Fox-Kean M (2005) Patent citations and the geography of knowledge spillovers: a reassessment. Am Econ Rev 95:450–460
2. Hendersen R, Jaffe A, Trajtenberg M (2005) Patent citations and the geography of knowledge spillovers: a reassessment: comment. Am Econ Rev 95(1):461–464
3. Li W, Holm E, Lindgren U (2009) Attractive vicinities. Popul Space Place 15(1):1–18
4. Hofstede G (1980) Culture's consequences: International differences in work-related values. Sage, Beverly Hills
5. Hofstede G (2001) Culture's consequences: comparing values, behaviors, institutions and organizations across nations, 2nd edn. Sage Publications, Thousand Oaks
6. Thomas DC (2002) Essentials of international management: a cross-cultural perspective. Sage, Thousand Oaks
7. Chaney LH, Martin JS (2004) Intercultural business communication, 3rd edn. Pearson Education Inc., Upper Saddle River
8. Asheim BT, Gertler M (2005) The geography of innovation—regional innovation systems. In: Fagerberg J, Mowery D, Nelson R (eds) The oxford handbook of innovation. Oxford University Press, Oxford, pp 291–317
9. Porter ME (2000) Locations, clusters, and company strategy. In: Clark GL, Feldman MP, Gertler MS (eds) The Oxford handbook of economic geography. Oxford University Press, New York, pp 253–274

10. Bjorkman I, Barner-Rasmussen W, Li L (2004) Managing knowledge transfer in MNCS: the impact of headquarters control mechanisms. J Int Bus Stud 35:443–455
11. Zander U, Kogut B (1996) Knowledge and the speed of the transfer and imitation of organizational capabilities: an empirical test. Organ Sci 6(1):76–92
12. Grant RM (1996) Prospering in dynamically-competitive environments: organizational capability as knowledge integration. Organ Sci 7(4):375–387
13. Spender JC (1996) Making knowledge the basis of a dynamic theory of the firm. Strateg Manag J 17(Winter Special Issue):45–62
14. Szulanski G (1996) Exploring internal stickiness: impediments to the transfer of best practice within the firm. Strateg Manag J 17(Winter Special Issue):27–43
15. Forsgren M, Johansson J, Sharma D (2000) Development of MNC centres of excellence. In: Holm U, Pedersen T (eds) The emergence and impact of MNC centres of excellence. Macmillan, London, pp 45–67
16. Hansen MT (1999) The search-transfer problem: the role of weak ties in sharing knowledge across organization subunits. Adm Sci Q 4(1):82–111
17. Allen T (1977) Managing the flow of technology: technology transfer and the dissemination of technological information within the R&D organization. MIT Press, Cambridge
18. Cohen WM, Levinthal DA (1990) Absorptive capacity: a new perspective on learning and innovation. Adm Sci Q 35(1):128–152
19. Nonaka I, Takeuchi H (1995) The knowledge creating company. Oxford University Press, New York
20. Gerybadze A, Reger G (1999) Globalization of R&D: recent changes in the management of innovation in transnational corporations. Res Policy 28:251–274
21. Mackinlay J (1986) Automating the design of graphical presentation of relational information. ACM Trans Graph 5(2):110–141
22. Storper M, Venables A (2003) Buzz: face-to-face contact and the urban economy. Paper presented at the. DRUID summer conference on creating, sharing and transferring knowledge: the role of geography, institutions and organizations. Copenhagen/Elsinore, 12–14 June 2003
23. Bathelt H, Malmberg A, Maskell P (2004) Clusters and knowledge: local buzz, global pipelines and the process of knowledge creation. Prog Hum Geogr 28(1):31–56
24. Kuemmerle W (1999) The drivers of foreign direct investment into research and development: an empirical investigation. J Int Bus Stud 30(1):1–24
25. Pauly LS, Reich S (1997) National structures and multinational corporate behavior: enduring differences in the age of globalization. Int Organ 51(1):1–30
26. Sawhney M, Prandelli E (2004) Communities of creation: managing distributed innovation in turbulent markets. In: Starkey K, Tempest S, McKinlay A (eds) How organizations learn. Managing the search for knowledge, Thomson Learning, London, pp 271–301

Chapter 19
Knowledge Transfer and Manufacturing

Abstract A manufacturing context will vary depending on a number of perspectives both internal and external. Many of these perspectives are difficult to influence, especially the business environment or competitive situation. However, the structural and infrastructural decisions could help a company to position itself in the competitive landscape in order to take advantage of business opportunities and become more adaptive. Structure is a premise provider for areas of knowledge that are most relevant and should be in focus particularly as a consequence of decisions regarding process technology. The location of facilities and vertical integration are structural decisions defining who will be key actors in knowledge creation, and hence can identify the challenges related to knowledge transfer. Infrastructure decisions are to a large extent about organizing and coordinating activities in a way that enables knowledge transfer. Infrastructure decisions are about how to obtain the best effect from the manufacturing structure. The knowledge dimensions of such decisions are reflected in day-to-day operations, and also knowledge creation for continuous improvement, and research and development projects. Structure and infrastructure decisions differ between manufacturing paradigms.

19.1 A Model

A manufacturing context will vary depending on a number of perspectives both internal and external. Many of these perspectives are difficult to influence, especially the business environment or competitive situation. However, the structural and infrastructural decisions could help a company to position itself in the competitive landscape in order to take advantage of business opportunities and become more adaptive.

A. Rolstadås et al., *Manufacturing Outsourcing*,
DOI: 10.1007/978-1-4471-2954-7_19, © Springer-Verlag London 2012

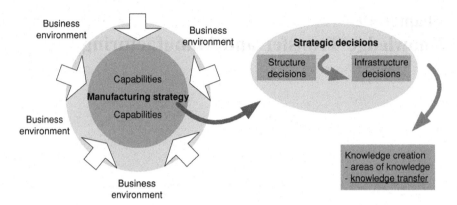

Fig. 19.1 Manufacturing strategy and knowledge transfer (Henriksen 2010) [1]

The relationships between the strategic context, structure and infrastructure decisions, and knowledge are summarized in Fig. 19.1, which shows the basic elements of a framework for knowledge transfer related to manufacturing strategy.

Structure is a premise provider for areas of knowledge that are most relevant and should be in focus within manufacturing, particularly as a consequence of decisions regarding process technology. The location of facilities and vertical integration are structural decisions defining what will be key actors in knowledge creation, and hence the challenges faced by knowledge transfer. Infrastructure decisions are to a large extent about organizing and coordinating activities in a way that enables knowledge transfer.

The knowledge dimension can be found in most strategic manufacturing decisions. The decisions about manufacturing structure regarding processes have implications for what kind of knowledge is needed to develop capabilities. Vertical integration, and the location, size, and roles of manufacturing units, all have implications for where to find knowledge and who should be involved in the knowledge creation, which in turn have major implications for knowledge transfer. Infrastructure decisions are to a large extent about how to obtain the best effect from the manufacturing structure. This includes decisions related to systems, organization, management, business culture, and so on. The coordination of operational, tactical, and strategic processes is the main concern of infrastructure decisions.

The knowledge dimensions of such decisions are reflected in day-to-day operations, and knowledge creation for continuous improvement, and research and development projects. Knowledge transfer becomes much more challenging when manufacturing satellites worldwide are expected to be premise providers. To what extent the strategic decisions are implemented and/or have the desired effects, needs to be measured. As the manufacturing strategies are increasingly focused on capabilities related to flexibility, customizations, innovations, sustainability and other less tangible aspects, the measurement will present challenges.

19.2 Paradigms and Knowledge

Even if a manufacturing strategy does not copy everything from a particular manufacturing paradigm, it will normally be influenced by practice, principles and beliefs reflected in certain paradigms. The paradigms represent guides for decisions on structure, for example supply relations and location issues, and in particular the role of the units. However, the paradigms will often be most exposed in decisions regarding organization, management, and culture.

An important dimension is whether a manufacturing strategy is based on a decentralized knowledge creation process, as in lean manufacturing, or a centralized knowledge creation process, as in mass manufacturing. The decentralized approach supports incremental innovations and continuous improvement, and is closely linked to tacit knowledge. A centralized approach is more relevant when the innovations are more radical and the knowledge more explicit. The knowledge transfer challenges will differ as tacit knowledge is made more explicit in order to be transferred to other units in the lean company. Transfer challenges related to explicit knowledge will be to make the knowledge relevant in a specific context or working environment.

Manufacturing paradigms have inherent principles and guidelines for how 'things get done', problem solving, stakeholder relations, and so on. This is typically reflected in innovation processes. Lundvall [2] distinguishes between interactive and linear models of innovation. A linear model has well-defined sequences and tasks within, for example, R&D projects. The linear model will often rely on analytical knowledge, while the interactive model is more relevant in incremental innovations. An interactive learning process between practitioners and experts (e.g. an R&D department) will normally rely on both a synthetic and an analytical knowledge base. The innovation process is much more decentralized, interactive and incremental within lean manufacturing than in mass manufacturing. More people, and different people, are involved in the lean knowledge creation process than in the more linear, formalized, and centralized innovation processes that we see in mass manufacturing.

The synthetic knowledge base is very important for innovations in lean manufacturing, whereas the centralized innovation process in mass manufacturing is often is more radical, depending on the analytical knowledge base. Craft manufacturing is extremely decentralized and dependent on the synthetic knowledge base, to some extent explained by its size, but the lack of standardization makes problem solving and incremental innovation a part of the 'daily routine'. Within adaptive manufacturing the innovation processes are often very complex, with companies dependent on profiting from common knowledge properties and innovations within networks. This requires continual improvements and incremental innovations, but also participation in radical innovations, as required within the network.

Based on the structural prerequisites, the innovation processes, behavioral aspects, etc., knowledge challenges and the relevant mechanisms and instruments for

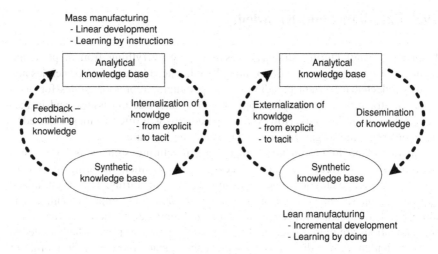

Fig. 19.2   Knowledge creation in mass- and lean manufacturing (Henriksen and Rolstadås 2010) [4]

knowledge coordination and transfer will differ between manufacturing paradigms. This is illustrated in Fig. 19.2 where mass and lean manufacturing are compared with reference to Nonaka's [3] knowledge creation spiral.

In Fig. 19.2 it can be seen that the spiral has different starting points and a different 'center of gravity' in lean and mass manufacturing. For example, the decentralized quality approach of lean manufacturing focuses on tacit knowledge and a main challenge is to make this knowledge explicit and useful for others. The centralized quality approach in mass manufacturing is based on explicit knowledge, where the challenges are more related to how to make people understand and implement quality principles and use appropriate tools. Differences in the knowledge creation processes are amplified when the mother plant—satellite context is introduced, especially when the distance, both cogitative and physical, increases between centralized knowledge creators and the people in the satellite units to which the knowledge is being transferred.

M.N. Kennedy, author of *Product Development for the Lean Enterprise* [5] focuses on tacit knowledge when he discusses the four cornerstones of the Toyota product development system. He describes how Toyota has created world-class cars without a formal development process, without administrative program managers, and without schedule slips. The product development is described as being all about consistent and planned knowledge from the subsystem level up (brakes, interiors, etc.), and not from the finished car down. The key is in a process that emphasizes "learning first". Toyota continually prototypes and tests to probe the limits of a given technology and the design tradeoffs associated with it. When General Motors (GM) wanted to learn the TPS, and established the NUMMI (New United Motor Manufacturing) plant in 1984 as a joint venture with Toyota, the NUMMI experience gave GM the necessary co-practice because the capabilities were "tacit know-how in action, embedded organizationally, systemic in inter-action and cultivated through learning by doing" [6, p. 570].

Even if there is little documentation in terms of manuals describing the development and improvement processes, in lean compared to mass manufacturing [7], there are many methods and principles which imply dealing with knowledge, such as the A3 problem-solving process. The A3 process has been described as a powerful method that systematically guides problem-solvers through a rigorous process, documents the key outcomes, and suggests improvements [8]. It is a tool used by Toyota to implement the PDCA [9] process.

Paradigms can be useful schemes for identifying knowledge aspects when they are described through a coherent set of criteria. A scheme by Jovane et al. [10] represents a useful framework for capturing knowledge aspects especially related to innovations, what kind of knowledge is relevant, and where to find it. In Table 19.1 some additional criteria are presented to further elucidates knowledge aspects.

19.3 The Knowledge Dimension of Sustainable Manufacturing

Sustainable manufacturing would normally represent a complex mixture of decisions and tradeoffs of goals and principles, which is also reflected in the knowledge area. This is illustrated in Fig. 19.3. Lean aspects, such as waste reduction and continuous improvement are important elements of sustainable manufacturing as they represent important means for economic efficiency and economic vitality. The implications for operations, innovations and improvements, and consequently knowledge resemble the ones previously presented for lean.

One of the challenges in making strategic decisions will be to make the tradeoffs represented by the intersections in Fig. 19.3. Decisions about processes and products will obviously have consequences for social progress and environmental protection. Choosing a manufacturing process that could damage the environment would violate the philosophy of sustainable manufacturing, even if it strictly from the company perspective could be the most efficient. In the same way supply chain decisions would be different when the supply chain partners' attitudes impacts on the environment are also brought to the table.

In sustainable manufacturing structural decisions about location of facilities could be made according to environmental issues such as energy efficiency, CO_2 emissions from transport, creating social progress in certain regions, protecting vulnerable environments, etc. This could be in contrast to a focus on optimized productivity and short term profit.

From these examples we see that a company basing its manufacturing on a sustainability strategy needs to capture and create knowledge from a wider field than in traditional manufacturing. These aspects are also different from the

Table 19.1 Manufacturing paradigms and knowledge

Aspect		Paradigm			
Field	Criteria	Craft manufacturing	Mass manufacturing	Lean manufacturing	Adaptive manufacturing
Literature sources		Womack Jones and Roos [7]	Ford (1926) Womack Jones and Roos [7]	Womack Jones and Roos [7]; Ohno [11]; Liker [12]; Kennedy (2003)	Pine [13]; Goldman, Nagel and Preiss [14]; Klocke [15]
Business model	Started	1850s	1910s	1980s	2000s
	Customer requirements	Customized products	Low cost products	Variety of products	Mass customized products
	Market	Pull Very small volume per product	Push Demand > Supply, Steady demand	Push–Pull Supply > Demand Smaller volume per product	Pull Globalization, segmentation Fluctuating demand
Innovations	Process enabler	Electricity Machine tools	Moving assembly line and DML	FMS Robots Modularized products	RMS Information technology
	Innovation process	Incremental	Linear and radical	Incremental and linear	Incremental and radical
	Behavior	Practical oriented skills (Learning by doing)	Centralized decision-making Learning by instructions	Decentralized decision-making. Continuous improvement. Learning by doing	Decentralized decision-making Knowledge to be applied instantly
Knowledge	Knowledge creation	Tacit knowledge	Explicit knowledge	Tacit knowledge	Tacit and explicit knowledge
	Knowledge base	Synthetic	Analytical	Analytical and synthetic	Analytical and synthetic
	Knowledge transfer/Challenge	Externalize knowledge communicating with customers	Internalize knowledge, for practical use	Externalize knowledge, making it more explicit	Continuously externalize and internalize knowledge
	Clustering of knowledge	Close to customers and craftsmen	Large units, not necessarily clustered	Close to customers and suppliers/network	Less spatially sticky, ICT as enabler for knowledge transfer

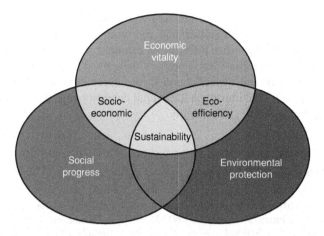

Fig. 19.3 Sustainable manufacturing

Table 19.2 Steps in the BEEM methodology (Henriksen and Røstad 2010) [16]

Business outcomes	Weight 1 = low 6 = high	Positioning in the market			Appl. target	Score	Weighted score	Risk of not achieving target
		Stay even 1	Significant advance 3	Technology breakthrough 9				

| Step 1 | Step 2 | Step 3 | | Step 4 | Step 5 | | Step 6 |

traditional knowledge fields in manufacturing—aspects that could be difficult to capture for a company. The knowledge creation process will also be different when the innovations are directed towards for example sustainable manufacturing processes, involving other people and resources than we might see in traditional manufacturing. The knowledge transfer issues are also often complex as knowledge related to social and environmental issues to a large extent are to be found locally. This context specific knowledge could be difficult to capture and transfer by head offices or units in other contexts. For example could local discussions on environmental issues be informal and have a tacit character. This tacit knowledge is an important dimension of sustainable manufacturing. However, to be able to deal with the complexity and tradeoffs illustrated we need means and methods that could simplify this picture. One way of doing this is to structure and formalize knowledge. BEEM (Business Effect Evaluation Methodology) is one example of such a method where the sustainability aspects might be counted for and weighted in the company's strategic decision process [11]. The major elements and steps in BEEM are shown in Table 19.2.

BEEM is addressing one of the challenges in a manufacturing strategy relying on initiatives and projects initiated and run from different units and how to evaluate, compare and prioritize them. If, for example, manufacturing satellites are given responsibility to initiate R&D projects there must be mechanisms to assure maximum objectivity and quantification in the decision process. These mechanisms should quantify the impact on the overall company objectives and strategies, in this example related to sustainability.

References

1. Henriksen B (2010) The knowledge dimension of manufacturing strategy: Mother plant-satellite manufacturing. Doctoral thesis NTNU, 2010:72. Trondheim, Norway
2. Lundvall BÅ (1992) National systems of innovation: towards a theory of innovation and interactive learning. Pinter Publishers, London
3. Nonaka I (1994) A dynamic theory of organizational knowledge creation. Organ Sci 5(1):14–37
4. Henriksen B, Rolstadås A (2010) Knowledge and manufacturing strategy—How different manufacturing paradigms have different requirements to knowledge. Examples from the automotive industry. Int J Prod Res 48(8):2413–2430
5. Kennedy MN (2003) Product development for the lean enterprise: why Toyota's system is four times more productive and how you can implement it. Oaklea Press, Richmond
6. Doz YL, Hamel G (1997) The use of alliances in implementing technology strategies. In: Tushman ML, Anderson P (eds) Managing strategic innovation and change. Oxford University Press, New York, pp 556–580
7. Womack JP, Jones DT, Roos D (1990) The machine that changed the world: The story of lean Production. Harper Business, New York
8. Sobek DK (1997) Principles that shape product development systems: A Toyota-Chrysler comparison, PhD Thesis. Ann Arbor: The University of Michigan
9. Kennedy MN (2010) Knowledge based product development—Understanding the true meaning of lean in product development. Presentation at the seminar *Knowledge Based Development Forum January* 27th–28th. Kongsberg, Norway
10. Jovane F, Koren Y, Boer CR (2003) A present and future of flexible automation: towards new paradigms. Ann CIRP 53(1):543–560
11. Ohno T (1988) Toyota production system: beyond large-scale production. Cambridge, MA: Productivity Press Inc.
12. Liker J (2004) The Toyota way: 14 managament principles from the world's greatest manufacturer. New York: McGraw-Hill
13. Pine BJ (1993) Mass-customization: the new frontier in business competition. Boston, MA: Harvard Business School
14. Goldman L, Nagel R, Preiss K (1995) Agile competitors and virtual organizations-Strategies for enriching the customer. New York: Van Nostrand Reinhold
15. Klocke F (2004) Adaptive manufacturing. The MANUFUTURE 2004 workshop, Fraeuenhofer IPT, Dortmund
16. Henriksen B, Røstad CC (2010) Evaluating and prioritizing projects—Setting targets. The business effect evaluation methodology BEEM. Int J Managing Projects Bus 3(2):275–291

Chapter 20
Outsourcing and Sustainability

Abstract Decisions on process technology have obvious consequences on sustainability and energy consumption particularly in regard to processes that need to energy efficient. Decisions related to supply chain integration also have consequences for example should a company that outsource its processes or chose certain supply chains (partners) be locked into a certain energy and sustainability policy. Structure decisions on where to locate facilities, their capacities and so on, can also have major consequences on sustainability. Infrastructure decisions have to be made accordingly to assure organizational and cross-organizational learning that can master sustainability issues. Implementing strategies based on sustainability can benefit from being incremental. The PDCA improvement approach can also be useful. Life cycle analysis is a methodology that is used to estimate and understand the environmental impacts of a product. Each phase of the life cycle—from materials extraction to end-of-life disposition—is ideally included in the assessment.

20.1 Energy Consumption on the Strategic Agenda

When sustainability and Corporate Social Responsibility (CSR) reach the strategy agenda in manufacturing companies a focus on energy consumption could be the right starting point. Energy consumption has clearly an environmental impact throughout the product lifecycle. Energy consumption is also an increasingly large part of the cost elements of manufacturing, thus energy efficiency and mastering energy consumption have substantial economic impact on companies. Figure 20.1 shows us how energy consumption is linked to sustainability indicators.

Decisions on process technology have obvious consequences on sustainability and energy consumption for example to which extent the processes are energy efficient. Decisions related to supply chain integration also have such consequences, for example, could certain supply chains (partners) be locked into an

A. Rolstadås et al., *Manufacturing Outsourcing*,
DOI: 10.1007/978-1-4471-2954-7_20, © Springer-Verlag London 2012

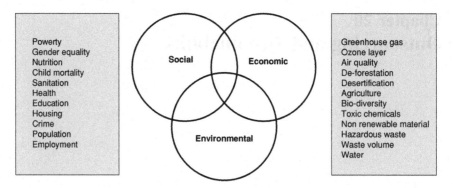

Fig. 20.1 Energy consumption in indicators for sustainability (Based on UN's indicators of sustainability) [1]

energy and sustainability policy. Structural decisions on where to locate facilities, their capacities, etc. also have major consequences on sustainability. However, infrastructural decisions have to be made accordingly to assure organizational and cross-organizational learning to master energy efficiency.

20.2 Energy Consumption: Process Technology

The environmental impact of energy consumption is important. Energy and cost savings related to manufacturing have been emphasized in practice, in improvement tools and management software, and a range of consultancy services. Governmental funding systems have also been supporting energy saving. This has resulted in the development of indicators used in for example quality systems, eco labeling and certificates, but also aggregated indicators at for example regional/national levels. Figure 20.2 illustrates how The European Commission has presented a climate package in a strategic triangle balanced between issues aiming at sustainable development, the improvement of European competitiveness, and maintaining security of energy supply [2].

Energy consumption indicators could be applied at all levels within industries even if they have initially been applied at national or regional level. Companies need indicators at strategic levels and indicators at a level of detail that could actually guide the company towards energy improvement. The World Energy Forum has presented four types of indicators for energy consumption:

- *Indicators of energy intensities to GDP*—value added (economic indicators in toe/USD) indicators of energy productivity
- *Indicators of unit consumption in physical units*—toe/tonne, toe or kWh/household, toe/employee, toe/vehicle, efficiency of power plants
- *Indicators of diffusion of energy efficient technologies* (e.g. cogeneration, electric steel, solar water heaters) *and practices* (e.g. share of public transport)
- *Indicators of CO_2 emission*—from energy combustion by sector.

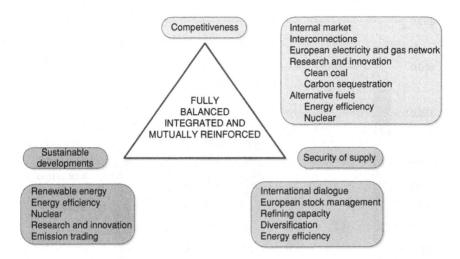

Fig. 20.2 Integrated climate and energy policy of the EU (Energy Efficiency Watch 2009) [2]

The figures below give a picture of energy consumption for one particular country, Norway. We see from Fig. 20.3 that there are big differences between sectors and that transport electricity intensive industries, and households count for the highest energy consumption. Figure 20.4 shows how energy efficiency has increased in terms of energy use/GDP. However, since GDP has increased considerably since 1990, energy use has also increased but by a lower percentage.

Even though energy costs don't count for more than approximately 2.5% of the total costs for a country like Norway, reduced energy consumption is important in most manufacturing industries. However, we see from Table 20.1 that the energy cost differs considerably between industries, where manufacturing of, pulp, paper products, and basic metals have the highest energy cost.

Industries can be classified into heavy energy consumer, medium energy consumer and low energy consumer. In Fig. 20.5 we see example of a low energy consumer—a vehicle assembly plant. Total energy costs in that example is equivalent to approximately 1% of the production output by the vehicle assembly plants, making it a relatively small cost factor in the total production process [5]. The example shows the distribution of electricity use in the different processes based on data from a number of plants in the US and Europe [6–8]. We see that there are big differences in energy consumption between the processes and we understand that process choice is important for energy consumption.

From the example of different construction materials in Fig. 20.6 we see that there are very big differences of energy costs ranging from stone (50 kw/m^{3}cubic meter) up to aluminum (141.500 Kilowatt per cubic meter) [9]. The picture gets much more complex when transport costs and other life cycle aspects of strategic decisions related to process technology and materials are included.

Implementing strategies based on sustainability and indicators of energy consumption could benefit from incremental improvement. The Lowell Center of

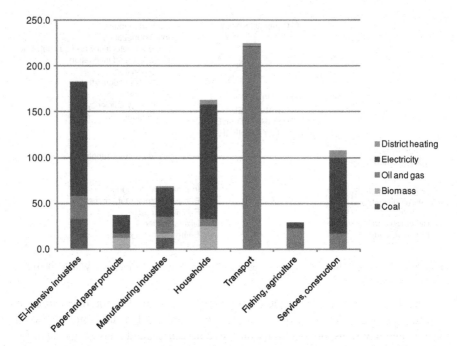

Fig. 20.3 Energy consumption by industry and source 2007 [4]

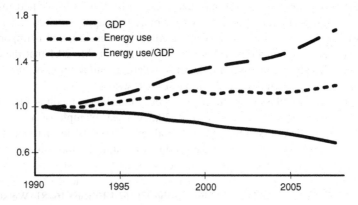

Fig. 20.4 Indicators for energy consumption, Norway [4]

Sustainable Production [10] has suggested five evolutionary levels where a company moving to a higher level of sustainability should include indicators from the new level, but also keep key indicators from previous levels:

1. *Facility compliance and conformance*—the first level measures to which extent the facility is in compliance with regulations and relevant standards

Table 20.1 Energy consumption by industry and Source, 2007 (SSB Statistics Norway 2009) [4]

	Total production cost	Energy consumption GWH	Energy cost	Energy cost % of production cost
Mining and quarrying	1,328	1,341	54	4.0
Manufacturing	75,203	68,247	1,961	2.6
Food products	16,890	4,555	261	1.5
Wood and wood products	2,609	1,672	53	2.0
Pulp, paper and paper products	1,700	8,792	191	11.3
Refined petroleum and chemicals	6,152	8,749	392	6.4
Non-metallic mineral products	3,985	3,137	125	3.1
Basic metals	7,886	24,071	700	8.9
Other	35,981	3,459	239	0.7

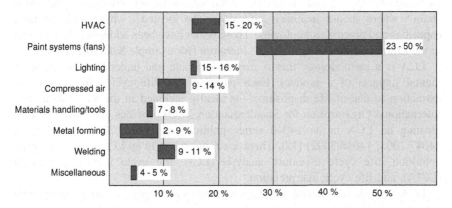

Fig. 20.5 Distribution of electricity in assembly vehicle plant [4, 5]

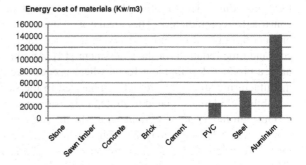

Fig. 20.6 Energy cost of various construction materials (Federation of Natural Stone Industries (SN Roc); CTBA, L'Essential sur le bois, 2001) [8]

2. *Facility material use and performance*—measure resource use efficiency and include measures of inputs, outputs and performance
3. *Facility effects*—measures the potential effect on environmental, worker and public health, community development and economic viability
4. *Supply chain and product Life Cycle*—promotes measurements similar to level 3, but includes supply chain, product distribution, use and ultimate disposal
5. *Sustainable systems*—show how the production processes fit into the larger picture of a sustainable society.

20.3 Outsourcing and Supply Chain Aspects

Even if sustainability could be implemented step by step, there is no doubt that we need to have perspectives outside the mother factory for a truly sustainable manufacturing strategy. This means that other structural decisions of the manufacturing strategy become important: what should be our position in the value chain?; where should facilities and partners be located?; what should be their capacities and process technologies? These issues have been addressed for example in the Life Cycle Assessment (LCA) literature (for example Santero et al. [11]).

LCA is a methodology that is used to estimate and understand the environmental impacts of a product. Each phase of the life cycle—from materials extraction to end-of-life disposition—is ideally included in the assessment. The International Organization for Standardization (ISO) provides guidelines for performing an LCA in its 14040 series publications (14040:2006; 14044:2006; 14047:2003; 14048:2002) [12]. There are four phases in an LCA; goal and scope definition, life cycle inventory analysis (LCI), life cycle impact assessment (LCIA), and life cycle interpretation.

LCA analysis is accepted and implemented as an important tool in sustainable manufacturing. However, implementing LCA for products has been criticized mainly from industry users as being too time consuming and costly [13]. The purpose of the simplified LCA is to address these criticisms of the LCA by simplifying the LCA procedures. The simplified LCA, however, contains two elements that could be construed as contradictory. They are the assessment of the environmental impact of a product throughout its entire life cycle with accuracy, and the minimization of the cost and time required for the assessment [14]. Another aspect of LCA is that they have a product development focus where process technology and other facets of manufacturing strategies are not emphasized in the same way. We also need sustainability guidelines related to strategic decisions i.e related to outsourcing.

Outsourcing is an important strategic decision in manufacturing, whereby a competitive advantage may be gained when products or services are produced more effectively and efficiently by outside suppliers. Kakabadse and Kakabadse [15] argue that the main reasons for outsourcing are: economic, quality, and innovation.

An additional argument for outsourcing could be to reduce risk related to unintentional environmental effects of operations. Through outsourcing of operations that are particularly exposed to such effects, the consequences of sanctions, penalties, litagation and so on, could be reduced for companies. However, this motivation for outsourcing has nothing to do with sustainable manufacturing and we argue for an outsourcing strategy that could strengthen sustainability through access to increased specialization, access to knowledge and innovation, and so on. In this perspective risk management should focus on reducing risk and negative consequences for sustainability from outsourcing. Acoording to Ravi Aron, these risks could be grouped into the following [16]:

- *Personalized, ubiquitous learning*
- *Operational risks*—possible slippages on quality, cost or speed of process execution
- *Strategic risks*—issues such as intellectual property, security and privacy
- *Composite risks*—longer term risks, such as losing the capability to execute such business processes in-house in the future due to loss of talent and knowledge of the business process.

There are several methods that could help decision makers in reducing outsourcing risks. There are also approaches and methods related to outsourcing that could support strategic decisions aimed at strengthening sustainability. These methods typically deal with partner selection and location decisions. The Analytical Hierarchy Process (AHP) [17] can be a relevant framework and has been popular among location consultants [18]. One of the challenges in applying these methods is to identify the relevant criteria describing sustainability for example energy consumption, and give these criteria the weight in line with their strategic importance.

Sustainability is a global issue and emphasize on energy saving stems to a large extent from the global need to save energy. This global need affects energy prices, emissions targets, and legislation. However, there are national and regional differences, and manufacturing companies should have a close look at energy consumption indicators to assure that outsourcing partners are selected, or localization decisions are made, in a way that meet the sustainability objectives.

For energy intensive industries and industries where energy costs are regarded as strategic issues, energy prices and access to energy could be major criteria for facility location or partner selection. In Fig. 20.7 we see other criteria that indicate that there are differences in energy intensity between nations and regions. For example we see that countries such as Russia and Ukraine (CIS) have the highest energy intensities. These differences could be explained by differences in energy prices that force companies and industries into energy savings, but could also be explained by different attitudes towards sustainability between governments (polices) and industries in these regions and nations. The differences in energy intensities adjusted for GDP indicates that there are differences in process technologies between regions that have effects on energy consumption, and hence energy costs.

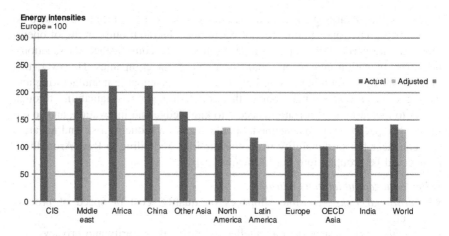

Fig. 20.7 Energy intensities, 2008: actual and adjusted at same GDP structure (Enerdata) [19]

Figure 20.8 shows national/regional differences in energy intensities for transport which measures the relation between transport energy use and the GDP. This could be an indicator of efficiency of the transport system, transport distances, but also the energy efficiency of the transport vehicles (e.g. trucks, trains, etc.). In other words this is also an important indicator for distribution challenges and costs in a region and consequently important in partner- and location decisions, when sustainability is emphasized.

20.4 Improvement in Energy Consumption

To be able to have a high level of control of energy consumption and continuously improve energy efficiency we need to focus on improvement processes and innovation. These aspects are again closely related to knowledge creation and transfer—organizational learning. As a consequence of these strategic decisions about localization and outsourcing, and choice of partners are not only a question of energy costs or energy efficiency today. The strategic decisions are also about which locations and partners could offer the best environment for continuous improvement and innovation that could give comparative advantages in the future. We also need to have mechanisms and methods that could enable and control those processes crossing organizational and national boundaries.

Life Cycle Analysis (LCA) is an important tool to bring facts about sustainability to the table. According to the ISO 14040 and 14044 standards, LCA is carried out in four distinct phases:

- Personalized ubiquitous learning
- Goal and scope definition

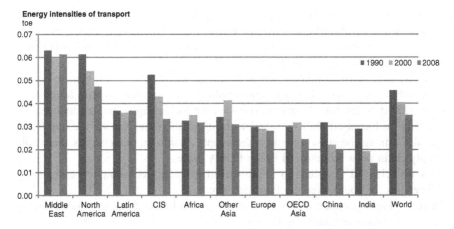

Fig. 20.8 Energy intensities of transport, 2008: actual and adjusted at same GDP structure (Enerdata) [19]

- Inventory analysis
- Impact assessment
- Interpretation.

Major corporations all over the world are either undertaking LCA in house or commissioning studies, while governments support the development of national databases to support LCA. A survey of LCA practitioners carried out by Cooper and Fava in 2006 [20] found that most LCAs are carried out with dedicated software packages and mostly used to support business strategy (18%) and R&D (18%), as input to product or process design (15%), in education (13%) and for labeling or product declarations (11%).

Simplified LCAs (SLCAs), are used to identify those areas which can be omitted or simplified without significantly affecting the overall results. Nevertheless, one aspect that is limiting the implementation and wide use of these simplified methods is their limited inclusion of cost and investment considerations, which excludes them of being used as key input for strategic decision making from a sustainability perspective. Another limiting aspect is the absence of a systematic method to select the relevant (from the environmental impact point of view) product attributes that would form the decision variables basis of product classification according to their critical life-cycle phase(s). An alternative could also be *Matrix methods* which are 'simple methods' to present as much information as possible about a product's environmental aspects in a systematic and clearly arranged manner. Commercially available web-based software could be used for rough analysis [21].

Reduced energy consumption and improved energy efficiency are closely related to energy management. Energy management and the process of monitoring, controlling, and conserving energy in a building or organization, is not surprisingly quite similar to the improvement processes, e.g. quality improvement that we find in manufacturing. Several stands, including ISO 14031, recommend that the indicators are developed and managed through the PDCA cycle. PDCA is a successive cycle which starts off small to test potential effects on processes, but then gradually leads to larger and more targeted change.

LCA and Matrix-methods are relevant in the PDCA improvement cycle, especially when product life cycle and supply chain issues are in focus. Muckstadt, et al. [22] describe how business processes are increasingly coupled and require not only an increased ability to deal with information and knowledge internally, but also among the stakeholders in the supply chains. In other words the whole supply chain must be involved in the PDCA cycle.

If we relate PDCA steps to improvement in energy consumption and energy efficiencies we could have the following process:

Plan: Finding opportunities to save energy, and estimating how much energy each opportunity could save.

Do: Taking action to target the opportunities to save energy (i.e. tackling the routine waste and replacing or upgrading the inefficient equipment). Typically you'd start with the best opportunities first.

Check: Tracking your progress by analyzing your meter data to see how well your energy-saving efforts have worked.

Act: You would typically analyze your meter data to find and quantify routine energy waste, and you might also investigate the energy savings that you could make by replacing equipment (e.g. lighting) or by upgrading your building's insulation.

20.5 Organizational: and Cross Organizational Learning

Energy management software (EMS) provides more concrete tools for reducing energy costs and consumption for buildings, processes or communities enabling the more concrete improvement-activities. EMS collects energy data and uses it for three main purposes: *reporting*, *monitoring* and *engagement*. Reporting may include verification of energy data, benchmarking, and setting high-level energy use reduction targets. Monitoring may include trend analysis and tracking energy consumption to identify cost-saving opportunities. Engagement can mean real-time responses (automated or manual), or the initiation of a dialogue between occupants and building managers to promote energy conservation. One engagement method that has recently gained popularity is the real-time energy consumption display available in web applications or an onsite energy dashboard/display.

However, the knowledge prerequisites for improvements in energy consumption are more than capturing and reporting facts e.g. through LCA- EMS (Energy Management Software) tools. Knowledge creation is a difficult process, involving people and their knowledge in a way that not only supports knowledge creation, but also accelerates it. Knowledge transfer is an imperative for knowledge issues, whether they are related to R&D or day-to-day operations. Argyris [23] has discussed the problem, focusing on extending cycles of learning from individual level to organizational level.

To improve company performance within sustainability e.g. related to energy consumption you must create knowledge, transfer and apply it not only within own organization but also involve manufacturing partners. Knowledge transfer requires a substantial amount of collateral knowledge on the part of the receiving organization to be able to decode and apply it.

One example is most buildings that have open to them a variety of equipment or building-fabric-related energy-saving opportunities, most of which require a more significant capital investment. Examples include upgrading insulation or replacing lighting equipment. Explicit knowledge represented by detailed power meter data won't necessarily help to find these energy saving opportunities. These kinds of energy saving projects will often be part of the tacit knowledge among workers at shop floor level or other people that see these opportunities in their daily work. This would to a large extent be tacit knowledge not captured in the fact based documentation and reporting systems. Consequently improvement in energy consumption is dependent on good systems and enablers for organizational- and cross- organizational learning. This process must include both the tacit and explicit knowledge.

However, it is much more reliable to base savings estimates on real metered data than on rules of thumb alone. It's also critically important to quantify the expected savings for any opportunity being considered. This means that energy indicators should be included in the information flows illustrated in supply chains and networks, in addition to LCA and the more dedicated EMS tools. But again, just finding the opportunities to save energy won't help save energy—action must be taken to target savings and this requires involvement and motivation from people at different levels within or between the organizations. Ideally the energy-improvement circle will be an ongoing effort to find new opportunities to target (Plan), to action (Do), and to track (Check) progress at making ongoing energy savings (Act).

When energy consumption becomes an important aspect of the manufacturing strategy this also requires a focus on partners and locations that could give the best opportunities for continuous improvements and innovation. This again requires a focus on infrastructure, organization, people, knowledge aspects, etc.

20.6 Sustainability at the Strategy Agenda

Sustainable manufacturing could be seen as the most recent manufacturing paradigm to emerge in the literature. Manufacturing companies defining themselves within this paradigm must deal with social and environmental aspects of their operations. However, companies that do not have sustainability as a part of their vision also need to deal with environmental and social aspects of their activities due to regulatory requirements. We have also seen that there could be considerable cost reduction potentials in focusing on environmental impacts Capturing the environmental aspects of manufacturing requires that they are an integrated part of the company strategies. Company objectives, goals and indicators must be developed, and implemented through decisions on structure and infrastructure.

The structural decisions represent the physical manifestation of the operations, such as process technology, location and capacity of facilities. These decisions all have important environmental facets exemplified through:

- Different energy consumption from process technologies (also a consequence of product technologies).
- Locations have different energy costs from; energy prices, transportation costs and energy efficiencies.

A third group of strategic structural decisions is the 'make-or-buy' decisions. These are decisions regarding the position in the value chain, and what the company should do themselves and what to outsource. We have seen that outsourcing has increasingly been seen as a way for manufacturing companies to focus their resources on what they define as core activity. Outsourcing will normally raise coordination and management issues for example related to improvement in sustainability.

Our arguments are that the infrastructural decisions enabling continuous improvement and innovations are of particular importance to master energy consumption and other sustainability aspects—maybe even more important than the structural decisions. The PDCA cycle represent a framework for improvement in energy consumption, which is closely related to knowledge and knowledge transfer. Manufacturing companies and supply chains need systems and approaches that enable quantification of data and knowledge, and there are several relevant LCA and EMS tools available for this. However, to get the involvement needed and collaboration with outsourced units, the tacit knowledge must also be captured. This means that the knowledge creation spiral of Nonaka and Takeuchi [24] is very relevant in sustainable manufacturing.

References

1. UN department of Economic and Social Affairs (2007) Indicators of sustainable development: Guidelines and Methodologies. http://www.un.org/esa/dsd/dsd_aofw_ind/ ind_index.shtml. Accessed Apr 2011
2. Koskimäki PL (2008) Energy efficiency policy and the importance of measurement and evaluation. Harmonized methods for evaluating energy end-use efficiency and energy services conference. Brussels, 15th Oct 2008
3. Energy Efficiency Watch (2007) Screening of National Energy Efficiency Action Plans. http://www.energyefficiencywatch.org/fileadmin/eew_documents/Documents/Results/EEW_ Screening_final.pdf. Accessed May 2011

4. SSB Statistics Norway (2009)
5. Galitsky C, Worrell E (2008) Energy efficiency improvement and cost saving opportunities for the vehicle assembly industry, LBNL-50939-Revision
6. Price A, Ross MH (1989) Reducing industrial electricity costs—An automotive case study. Electricity J 2(6):40–51
7. Dag S (2000) Volvo faces a deregulated european electricity market. Department of Mechanical Engineering, Linköpings Universitet, Linköping, Sweden
8. Leven B, Weber C (2001) Energy efficiency in innovative industries: application and benefits of energy indicators in the automobile industry. In: 2001 ACEEE summer study on energy efficiency in industry proceedings Vol 1. American Council for an Energy-Efficient Economy (ACEEE), Washington, D.C. pp 67–75
9. Monde Diplomatique (2007) Atlas Environment du Monde Diplomatique
10. Veleva V, Ellenbecker M (2001) Indicators of sustainable production: framework and methodology. J Clean Prod 9(6):519–549
11. Santero N, Masanet E, Horvath A (2010) Life-cycle assessment of pavements: a critical review of existing literature and research, SN3119a, Portland Cement Association, Skokie, Illinois
12. International Organization for Standardization (2006) Environmental management—Life cycle assessment—Principles and framework. ISO 14040:2006(E)
13. Christiansen K (1997) Simplifying LCA: Just a cut? Final report from the SETAC-EUROPE LCA screening and streamlining working group (Brussels: SETAC-Europe) from the SETAC-EUROPE LCA Screening
14. Stoeglehner G, Narodoslawsky M (2008) Implementing ecological footprinting in decision-making processes. Land Use Policy 25(3):421–431
15. Kakabadse N, Kakabadse A (2000) Critical review—outsourcing: a paradigm shift. J Manag Dev 19(8):670–728
16. Squidoo (2011) Reducing operational risk in business process outsourcing. http://www.squidoo.com/operational_risk. Accesed Aug 2011
17. Saaty TL (1980) The analytic hierarchy process, planning, piority setting, resource allocation. McGraw-Hill, New York
18. Viswanadham N, Kameshwaran S (2007) A decision framework for location selection in global supply chains. IEEE international conference on automation science and engineering, CASE 2007
19. World Energy Council (2008) Energy efficiency policies around the world: review and evaluation—Executive summary. www.worldenergy.org. Accessed Mar 2011]
20. Cooper JS, Fava J (2006) Life cycle assessment practitioner survey: summary of results. J Ind Ecol 10(4):10–12
21. Wimmer W (2002) ECODESIGN pilot-product investigation. Learning and Optimization. Tool for Sustainable Product Development (2011) http://www.ecodesign.at/pilot/. Accessed Apr 2011
22. Muckstadt JA, Murray DH, Rappold JA, Collins DE (2001) Guidelines for collaborative supply chain system design and operations. Inf Syst Front 3(4):427–453
23. Argyris C (1990) Overcoming organizational defenses. Facilitating organizational learning. Allyn and Bacon, Boston
24. Nonaka I, Takeuchi H (1995) The knowledge creating company. Oxford University Press, New York

Part V
Cases

In this part of the book three cases from different industries and different manufacturing contexts are presented to highlight different knowledge aspects of outsourcing. The aim of these chapters is to illustrate challenges related to knowledge transfer that companies might face, but also to show means and enablers for knowledge transfer.

Chapter 21
Supply Chain Integration and Knowledge Transfer—A Case from the Automotive Industry (Case 1)

Abstract The automotive industry has been important for developing new manufacturing principles and paradigms. The quality movement, lean manufacturing are examples. Automotive companies often rely on long and complex supply chains. In this example we see how a supplier in the automotive industry is integrated into the knowledge transfer process. We also see how tools and different methods can help to overcome some of the challenges in supply chain collaboration.

The automotive industry has been important for developing new manufacturing principles and paradigms. The quality movement, lean manufacturing are just two examples of major paradigms to emerge from this sector and which have now widespread adoption across all sectors of industry. Automotive companies often rely on long and complex supply chains. In this example we see how a supplier in the automotive industry supply chain is integrated into the knowledge transfer process. We also see how tools and different methods can help to overcome some of the challenges in supply chain collaboration.

21.1 The Strategic Context

The case company is among the 100 largest suppliers in the automotive industry. The revenues of the group were 905 million Euro in 2008 and they employ 9,000 people. With its headquarters in Scandinavia the company has almost 50 facilities in 20 countries providing system solutions to vehicle makers around the world. This example is concerned with one business unit within the truck market (driveline products) and has the characteristics described in Fig. 21.1.

The theory of lean manufacturing characterizes the role of suppliers, since lean is adapted widely within the automotive supply chains. The case company defines lean through manufacturing principles such as JIT, waste reduction, quality,

A. Rolstadås et al., *Manufacturing Outsourcing*,
DOI: 10.1007/978-1-4471-2954-7_21, © Springer-Verlag London 2012

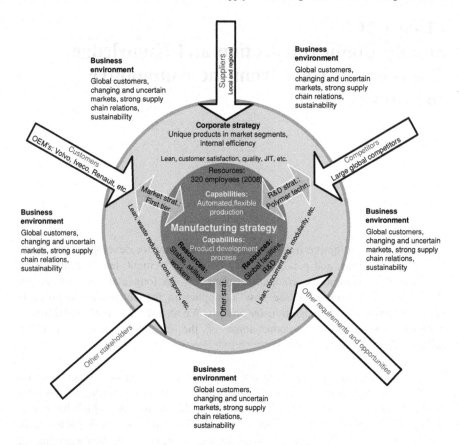

Fig. 21.1 The strategic context for supplier in truck manufacturing

customer orientation, modularization, concurrent engineering, and incremental innovations. The company has established long-term relationships with the leading OEMs in the commercial vehicle industry. This has given the company and the business unit a good knowledge of the industry and basis for understanding of expectations. The company is constantly evolving, in line with the industry's demand for continuous improvement, and increased functionality and effectiveness (3–5% increase productivity per year). Since 1995, the business unit has emphasized innovations and R&D programs. Today, the company's products meet all applicable industry standards and are optimized for minimum pressure loss, temperature independent sealing, compact external dimensions, and cost efficiency. The business unit has developed capabilities that are important for winning orders in the automotive industry. These capabilities are described by management as:

• Personalized ubiquitous learning
• Stable, cost-effective processes with zero defects

- Processes that comply with the customers JIT requirements
- Automated and flexible manufacturing processes adapted to customers' changing demands
- A product development process focusing on the customers' technology and product expectations for modularity, easy installation and quality requirements.

The company and the business unit have defined a mother plant—satellite manufacturing strategy. The mother plant has important roles within the R&D function. The mother plant has the main resources, not only in R&D, but also for serving global customers when it comes to product adjustments and major product improvements. The mother plant has a focus on the core product of the future, while the satellites produce less advanced products. The mother plant is particularly important for designing systems for manufacturing, and developing state of the art process technology. In line with lean principles, the satellites focus on quality and process improvement which often means improvement to the manual processes.

The business is in a strong growth phase with new manufacturing satellites now being planned. When these planned satellites are established in, for example, India and the USA they are expected to develop into 'server'or contributor factories [1], with responsibility for price policies, supplier selection, and also more tactical processes, such as local strategies and modifications, and even process and product system development. This implies a knowledge creation process, with knowledge transfer between the satellites.

21.2 Business Systems and Types of Supply Chain Relations

According to Stock and Lambert [2] supply chains consist of five interconnected business systems:

- *Management systems*—planning (strategic and tactical), measurement and reward
- *Marketing systems*—product, price, promotion, place (how, where, how much)
- *Engineering systems*—product, design, process design
- *Manufacturing systems*—production capacity, production scheduling, shop floor decisions
- *Logistics systems*—place, inventory, transportation, sourcing.

The level of integration between companies for each of these business systems depends on how processes, information systems and decision systems are integrated [3]. Based on the integration levels Muckstadt et al. [3] define four types of supply chain relationships:

1. *Communicators*—companies with basic integration. Customers place orders and the company is expected to deliver within a requested lead-time. The level of information systems- and business process integration is low

2. *Coordinators*—such companies capture and share more detailed operational data, but decision systems are not integrated at strategic or tactical levels. Business processes are normally not highly integrated
3. *Co-operators*—supply chain relationships where customers share information to the extent that they communicate plans that are out of the ordinary. This require both integrated information infrastructure and supporting business processes
4. *Collaborators*—companies work together at strategic and tactical levels. They execute collaboratively to achieve the maximum system effectiveness. Information systems and business processes are highly integrated.

All the above relationships might be present at the same time within a supply chain. Integration of supply chains will often develop over time from fragmented supply chains mainly focusing on integration of steps in internal supply chains (communicators), to coordination of activities between businesses. The final phases are a 'real time' synchronization and coordination of planning activities across the supply chain, and finally partnership and strategic alliance arrangements. The principles underlying these changes are the recognition that supply chains are the basis for today's competitiveness [4].

21.3 Integration in the Truck Supply Chain

In the dynamic automotive business environment, the supply chain is one critical element helping automakers to differentiate themselves from the competition, and trends are reinforcing the need to redefine supply chain strategies, layouts, and operations. The evolution and transformation of technologies, market structure, and customer needs are pushing firms to reduce the time to market of products, increase technological and organizational flexibility, and gain access to available local and global knowledge and capabilities [5]. Global first-tier suppliers are normally expected to take major responsibilities in the development of global product platforms, and process improvements for JIT and continuous cost reductions [6]. Zagnoli and Pagono [5] summarize the implications of implementing lean principles for suppliers in the automotive industry as follows:

• The absolute emphasis by OEM, on market orientation and customer satisfaction
• The growing relevance of modular product architecture as opposed to integrated architecture
• New organization of internal and external processes, leading to internal and external networks.

Figure 21.2 shows how organizations are organized into first tier, second tier, and so on suppliers and where the distribution of responsibilities requires supply chain collaboration.

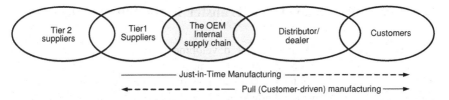

Fig. 21.2 Truck supply chain—collaboration along the supply chain

Table 21.1 Integration the truck OEM—supplier

Business system	Business processes		Information systems		Decision systems	
	High	Low	High	Low	High	Low
Management	X		X			X
Marketing	X		X			X
Engineering	X		X		X	
Manufacturing	X		X		X	
Logistics	X		X		X	

There is a high level of vertical integration in the supply chain and also highly integrated R&D and improvement processes. This is the key to creating knowledge for developing capabilities of efficient, flexible, and customer-oriented processes. These capabilities apply to the role as a first-tier supplier in the automotive industry.

Table 21.1 shows an assessment of the integration between the truck OEM and the supplier. A high level of integration in found among almost all business systems between the OEM and the Tier 1 supplier's. The integration of business processes and information systems are to a large extent explained by general industrial standards and requirements in the automotive industry such as the quality system (ISO/TS 16949:2002).

Even though the decision systems can be characterized as integrated they do not seem completely balanced. Lean principles motivate suppliers, especially first tier. However, the OEM is taking most major decisions and is defining the basic requirements for new products. OEM's are also putting pressure on the whole supply chain for continuous improvement. Since the overall quality management systems are focused on the OEM information systems there are definitely elements of communicator-relationships between the OEM and the suppliers. However, major responsibilities for product development and quality improvement are addressed to the first tier suppliers and the relationship between the supplier and the OEM can be characterized as 'collaborators'. This is underlined by the fact that since the numbers of direct suppliers (first tier) are reduced, the ones that remain have increased responsibilities towards both the sub-suppliers (second tiers, etc.) and the OEM.

21.4 Fact Based Knowledge Transfer in the Supply Chain

Continuous learning and incremental development are basic elements of lean manufacturing [7], and has a big impact on improvement and quality approaches in the truck manufacturing industry. The way lean puts employees to the forefront of improvements based on their knowledge, has similarities to what we see in for example in craft manufacturing. However, there is a major difference since lean manufacturing aims to base improvements on facts and not only the tacit knowledge. In this case from truck manufacturing the quality improvement is largely based on explicit knowledge, where integrated processes require that quality improvements and changes are based on facts. Facts that are often made available through integrated information systems.

This case study has many of the characteristics from lean manufacturing described in the Toyota Production System (TPS). Knowledge of how to make Toyota cars lies embedded in highly specialized social and organizational relationships that have evolved through decades of common effort [8], but the knowledge transfer and formalization of tacit knowledge is enabled in information systems, routines, and robust processes. There is a risk for not capturing all the necessary knowledge aspects in this conversion. In this case study the knowledge requirements defined by the OEM are quite strict and formal. Since they are related to some key quality measures there are risks that some of the informal contact and transfer of tacit knowledge is neglected. However, this risk will normally be much higher in mass manufacturing, where the formal planning and measurement systems are even more strict and top-down oriented [9].

The level of integration of the supply chain is a prerequisite for knowledge transfer. The way business processes, decision systems and information systems are integrated tells much about the type of, quantity and format of the knowledge needed to transfer it along the supply chain. Quality improvement in truck manufacturing is more fact oriented, based on explicit knowledge where integrated supply chains enable knowledge transfer and collaboration around quality improvement. Even though tacit knowledge plays a more important role in lean manufacturing than in mass manufacturing, there is also a risk that the integrated structure and systems in lean manufacturing could over-emphasize explicit knowledge. This knowledge often fits into systems where tacit knowledge and personal relations might be overlooked.

References

1. Ferdows K (1997) Making the most of foreign factories. Harward Bus Rev 75:73–88 (March–April)
2. Stock JR, Lambert D (1997) Strategic logistics management, 2nd edn. Irwin, IL
3. Muckstadt JA, Murray DH, Rappold JA, Collins DE (2001) Guidelines for collaborative supply chain system design and operations. Inf Syst Front 3(4):427–453

4. Hill T (2000) Operations management: strategic context and managerial analysis. Macmillan, London
5. Zagnoli P, Pagono A (2001) Modularization, knowledge management and supply chain relations: The trajectory of a European commercial vehicle asseler. Actes du Gerpisa 32:45–64
6. Morrell J, Swiecki BF (2001) E-readiness of the automotive supply chain, just how wired is the supplier sector? ERIM, Center for Automotive Research, Center for Electronic Commerce Available at http://www.cargroup.org. Accessed April 2007
7. Liker J (2004) The Toyota way: 14 managament principles from the world's greatest manufacturer. McGraw-Hill, New York
8. Bandaracco JL (1991) The knowledge link: how firms compete through strategic alliances. Harvard Business School Press, Boston
9. Henriksen B, Rolstadås A (2010) Knowledge and manufacturing strategy—how different manufacturing paradigms have different requirements to knowledge. Examples from the automotive industry. Int J Prod Res 48(8):2413–2430

Chapter 22
Quality Improvement in Craft Manufacturing—A Case from Leisure Boat Manufacturing (Case 2)

Abstract Competition requires that craft manufacturers and their supply chains innovate, improve, and increase their efficiency to meet the challenges from globalization and other forces for change. The leisure boat industry is an example where manufacturers are facing increased competition. 'High end' leisure boat manufacturers are now grappling with the challenge of how to meet this competitive pressure while preserving and improving their unique quality of craft manufacturing.

Competition requires that craft manufacturers and their supply chains need to innovate, improve, and increase their efficiency to meet the challenges from globalization. The leisure boat industry and other 'high end' leisure boat manufacturers are now grappling with the challenge of how to meet this competitive pressure while preserving to improve their unique quality in craft manufacturing.

22.1 The Strategic Context

The leisure boat industry is characterized by small and medium sized enterprises competing on the global scale. Traditionally these companies have been small but the last two decades there has been a tendency towards larger more industrialized companies with higher volumes and reduced costs. This has changed the industrial structure, increased competition and challenged traditional craft manufacturing principles. The strategic context for the leisure boat manufacturer (OEM) is shown in Fig. 22.1.

The OEM in this case study has a turnover of 25 million Euro (2010), employs 100 people (2010), has manufacturing satellites in two countries in addition to the mother plant, and has customers and suppliers on all continents. Even though the company has a defined product line with boat models from 25 to 58 ft, the level of customization is high making the manufacturing processes difficult to automate. For instance, the customer's choice of a larger water-tank

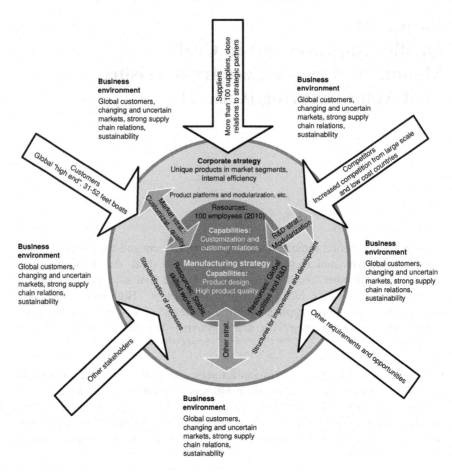

Fig. 22.1 The strategic context for leisure boat manufacturer (OEM)

impacts on the boats stability, thus requiring a reconfiguration of the instalment of some components and numerous other adjustments to brackets and bars. The manufacturing processes are mainly manual and adjustments need to be made regularly to all stages of the boat manufacturing process. The quality of the OEM's boats is perceived by the market to be very high. In summary, the company has many of the characteristics of craft manufacturing industry.

Even though the OEM has taken steps towards integration of its internal supply chain, it has a functionally oriented structure, with several built-in delays and inventories. There are very few formal reporting structures and systems, and those that do exist are to some extent process-oriented. Nevertheless, people in the manufacturing process do communicate with the technical offices in planning and problem solving.

Each boat is customized and the customer is involved during the whole manufacturing process and can dictate changes at all stages of the process. However, the relationship between the customer and the manufacturer (OEM) is not necessarily highly collaborative. Rather, the customer often dictates changes the manufacturer simply has to adjust to them. This often leads to interruptions and delays, with reduced margins as a result.

22.2 Quality Improvement as an Integrated Part of Craft Manufacturing

One of the main advantages in craft manufacturing is that the employees have a strong ownership to their products and put a lot of effort in finding good ad-hoc solutions to problems. For example, when a supplier has delivered a part with minor quality problems the craftsmen can often fix the problems themselves instead of initiating administrative processes. The craftsman represents a capability for making adjustments and practical solutions in line with customer expectations. A general characteristic of quality and quality improvement in the leisure boat company is that it has employees constantly work to optimize quality.

The PDCA Cycle is considered as a good description of how to work on quality improvement in most industries. As previously described the cycle consists of four defined and continuously repetitive steps:

Plan—Design or revise business process components to improve results
Do—Implement the plan and measure its performance
Check—Assess the measurements and report the results to decision makers
Act—Decide on changes needed to improve the process.

The PDCA Cycle [1] is normative as it describes steps that should be taken for deliberate continuous improvement, but could also be used as a reference for most quality improvements even for companies that are not following a structured quality improvement approach.

In Table 22.1 quality improvement in the leisure boat example is described according to the PDCA improvement cycle.

Quality improvements in the leisure boat supply chain are most often initiated from experiences and problem solving related to daily work/operations, but also from issues raised by customers and suppliers. This is often done through informal communication, personal relations, rather than on formalized procedures or well defined performance requirements (tangible measures) that were seen in the truck supply chain case study earlier.

In the leisure boat case decisions on full scale implementations are normally made through discussions and evaluations by operators and other people concerned by the changes. Also knowledge workers from the OEM can be involved in the decision process, that normally is informal and where changes are not well

Table 22.1 PDCA-cycle in leisure boat manufacturing

Steps	Description of Leisure boat manufacturing
Plan	There are quality systems, product descriptions and manuals, but few process descriptions at shop floor-level and between companies. This implies that the performance measurement systems are absent or not capable of indicating the need for improvements
	Changes are often based on discussions and personal experience. The "red books" used by craftsmen is an important tool for quality improvement. The improvements are not communicated or registered anywhere else than in the suppliers book
Do	Planned changes are often easy to implement in a small-scale situation. The people behind the proposed changes are testing them in their daily operations. Even though the rationale for the changes are not formally described, these workers have knowledge about why and how changes are to be made
Check	The processes are not described in details and the performance measurement systems are poorly developed
	When the effect of the changes are evaluated and decisions about full implementation are made they are often based on experiences and discussions among the people involved in the processes rather than systematic facts
Act	The lack of standardized and described processes gives challenges to implementation. Involving several operators in the change process would make the implementation easier. When operators are replaced, their knowledge about the changes (improvement) is often lost

documented. This decision process can add knowledge aspects (tacit knowledge) that are not necessarily covered by the formal decision processes.

The involvement and ownership among the operators, or craftsman, could be one important explanation for the survival of craft manufacturing companies. However, the tacit character of knowledge could also be a drawback for improvement and represents a conserving element. This is illustrated in statements such as "we have made boats for decades and know how to do this". This is amplified by the lack of documentation and facts that could represent more generally accepted arguments for improvements and change.

22.3 Integrating Suppliers into Quality Improvement

The suppliers are to some extent decoupled from the OEM. Contacts between the OEM and its suppliers are often ad hoc and based on personal relations. Among the OEM's more than 1,000 suppliers, some are of strategic importance (e.g. suppliers of engines). However, for most suppliers the OEM is a small customer with little leverage. In other words, the supply chain power of the OEM is limited. This is illustrated by one supplier case company, which has gained a monopoly position in its region. This supplier has a turnover of 10 million Euro and employs approximately 70 people (2010). The company is located in an agglomeration of leisure boat manufacturers, and develops and produces customized products (often in low volumes) to a large number of boat manufacturer's (OEM's). The supplier is recognized for high quality products, and has a competitive advantage that has prevented the entrance of competitors. The manufacturing processes are to a large extent manual but aided by low technology machine tools. Integration with the OEMs varies, with some interaction at the design phase, and less in the manufacturing phase. In general, the supplier is decoupled from its customers in the manufacturing phase, except from contact pertaining to receiving orders and problem solving related to quality issues. A typical situation is the difficulty to define when a product has the right quality due to poorly documented processes. The two companies have agreed on a system to handle claims from final customers and a higher level of integration of decision systems (management system).

Decision systems for the supplier are quite integrated with the OEM within engineering and manufacturing. This is based on informal and ad hoc personal contact rather than formalized information systems and processes. Even though the supplier and the OEM are using the same manufacturing planning and control systems, they are not integrated.

In spite of little or no interaction between the supplier and the OEM within important areas, we can characterize the relationship as *communicators* [2]. The supplier receives forecasts and orders from the OEM by e-mail and telephone, but there is low integration of business processes. Moreover, as the supplier has more customers than this OEM studied, they have short-term flexibility to juggle orders according to their own needs. This is not necessarily aligned with the needs of the supply chain.

22.4 Infrastructure for Knowledge Transfer

In leisure boat manufacturing high quality products are achieved through employee ownership of products and processes, and their responsibility for quality and quality improvement. They do so by discussing tacit knowledge together. Challenges appear when quality issues should be discussed and communicated between people operating in different contexts, for example between supplier and OEM. Quality improvement in craft manufacturing is very dependent on the individual skills and initiatives from the employees involved in the process. However, the leisure boat manufacturing process has identified a potential for a better organization of work on quality improvement, but also for knowledge transfer in general. The company wants to standardize more of their manufacturing processes and to identify and share best practice.

This was achieved by introducing a more formal and explicit model for knowledge transfer, defining project oriented teams as the basic organizational element, building arenas for knowledge transfer (meetings) and finally formalizing the knowledge transfer e.g. through a defined format for improvement suggestions

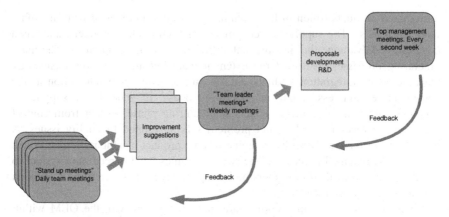

Fig. 22.2 A more structured way of knowledge transfer in leisure boat company

and the more comprehensive proposals being communicated to top management. The model is illustrated in Fig. 22.2.

Daily team meetings (maximum 10 min) are conducted every morning and where employees are expected to discuss concrete ideas for improvement. Some of the suggestions are so simple and with very few implications for others that they can be agreed upon and implemented immediately. However, there are other suggestions that need to be discussed with other teams and team leaders. The weekly team leader meetings require a more formal (but still relatively simple) process. These meetings have the authority to take decisions within certain resource limits, but the most comprehensive suggestions need to be described in a standardized format and discussed by the top management in their formal meetings.

The feedback system of this model is very important to make this infrastructure as the preferred way for the employees to involve themselves in improvement. Whatever the outcome or results of suggestions and proposals are, a standardized and rapid feedback to the people that has initiated ideas is required. In addition, suppliers are included in this model through their contact with team leaders and direct contacts with the craftsmen) The model has shown the importance of discussing quality issues with colleagues in a more formal structure, the stand-up meetings every day and tacit knowledge transfer being transferred. The model also means that knowledge is made more explicit when it is brought to a higher level in the organization.

References

1. Deming WE (1986) Out of crisis Boston. MIT/CAES, Cambridge
2. Muckstadt JA, Murray DH, Rappold JA, Collins DE (2001) Guidelines for collaborative supply chain system design and operations. Inf Syst Front 3(4):427–453

Chapter 23
Adaptive Manufacturing and Real Time Knowledge—A Case from Furniture Manufacturing (Case 3)

Abstract In the race for competitiveness and cost reduction, more automated processes are seen as an important approach. One of the challenges is to create automated processes that keep the flexibility of more manual processes. In this case we see how this is done in a company within the furniture industry. The case illustrates how the company gets access to data that creates a basis for increased knowledge but also where attention has to be put on the human dimension.

In the race for competitiveness and cost reduction more automated processes are often seen as a strategic solution. One of the challenges has been to create automated processes that maintain the flexibility of more manual processes. In this case we see how this is done in a company within the furniture industry. The case illustrates how the company gets access to data that creates a basis for increased knowledge but in addition also pays attention to the human side.

23.1 The Strategic Context

This European case develops, manufactures, markets and sells furniture and mattresses. Sales are essentially aimed at the market for home furnishings, although sales are also made within the contract market. The company is a supplier of branded goods. The Group had total sales revenues of 400 million Euro and a total workforce of more than 1,600 in 2010. The strategic context is shown in Fig. 23.1. Figure 23.2 shows operating revenues from 2001 to 2010.

The market segments wherein the company operates is in the high end where branding combined with functionality are important for winning customers. However, quality/price is also important. Expertise in brands and brand building and an international marketing concept are key elements in the company's operations. The furniture company defines itself as a competence-driven company, which makes extensive use of modern and advanced production equipment.

A. Rolstadås et al., *Manufacturing Outsourcing*,
DOI: 10.1007/978-1-4471-2954-7_23, © Springer-Verlag London 2012

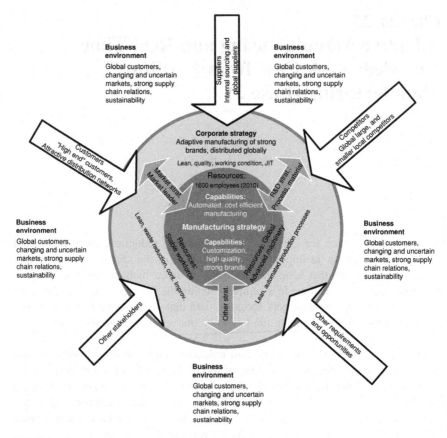

Fig. 23.1 The strategic context for furniture manufacturer

This includes a high degree of automation in the manufacturing facilities and automated internal transport of components.

In 2004 the company opened a new 30000 square meter plant, employing almost 50% of the workforce. This included state of the art technology in automated processes. The company also has production in 5 other plants in the region. These plants are to a large extent specialized and focused on optimized and flexible processes for defined products. The smaller plants have also continuously invested in advanced production processes. The company has in-sourced much of the production of components and in particular parts of the production that have been automated, with extensive use of robots. For example one plant that manufactures foam and springs, has a separate quilting unit and also produces its own wooden frames. This means that it controls the whole production process and the quality of the mattresses. These factories also produce components for other companies.

The reason for a strategy based on principles and technologies for adaptive manufacturing is the common understanding of how to meet the challenges faced

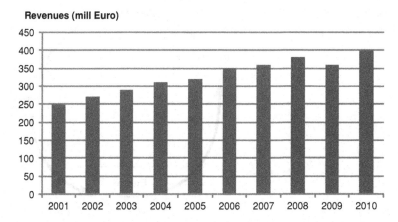

Fig. 23.2 Development in operating revenues

by competition in low cost countries. The investments made have been extensive, and the effects on productivity and sales have been positive even in the difficult market situation from 2008. The company also has a strategy of producing using plants in the nearby region and not setting up units in different geographical markets or in low cost countries. The company believes that increased competitiveness is gained through high tech adaptive manufacturing with good control over processes, products and technology.

23.2   The Three Elements of Adaptive Manufacturing

An adaptive factory can be viewed as a manufacturing system that can adapt extremely fast to changes in products, constraints, customer and market requirements, and sales. The focus is on flexibility in the short and long term, both on product variants and volume. This is supported by Maier-Speredelozzi et al. [1] stating that with increased consumer demands for a wider variety of products in changeable unpredicted quantities, manufacturing system responsiveness has become increasingly important for industry competitiveness.

Adaptive manufacturing can be defined as the continuous automatic adaptation of manufacturing resources and production processes to their evolving environment. It overcomes existing limitations through intelligent combinations of innovative processes, and handles the transfer of manufacturing know-how into totally new manufacturing-related methods. The basic idea of adaptive manufacturing is to create processes that are self-adjustable through: measurement, deciding on actions based on collected data, and execution of decided actions. The aim is to create more robust processes that can handle changing input and conditions, and still produce good results. According to SAP [2], adaptability has two primary characteristics, flexibility and velocity. Flexibility enables a manufacturing

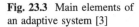

Fig. 23.3 Main elements of
an adaptive system [3]

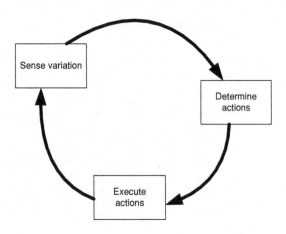

unit to scale efficiently while velocity determines its ability to switch operational modes rapidly and to transition between modes such as high-volume/low-mix to high-volume/high-mix product loadings.

Manufacturing systems are becoming more module-based. Larger units like five-axis milling machines, are constructed and assembled by mechatronic modules which can be swiftly exchanged and/or reconfigured. The mechatronic modules become more intelligent through embedded software and sensors which enables the modules to become autodidact, self-adjustable and self-repairable. Man–machine interfaces become more standardized. Figure 23.3 shows the main elements of the adaptive manufacturing system of the furniture case company.

Adaptive manufacturing concerns the three elements, sense variations, determine actions, and execute actions [3]:

- To sense variations automatic measurement systems such as, vision systems, temperature sensors, vibration sensors, etc. are common, but also data about market developments, customer requirements, financial data, etc. are part of the data collection for adaptive manufacturing
- Based on the data and information gathered adaptive manufacturing needs decision support systems and decision processes, etc. but also human interpretation
- The third element is to automatically adjust e.g. speed/temperature, etc. in processes and improve the processes. However, the improvement processes will often require organizational changes or changes in peoples' attitudes, behavior, etc.

Machine vision is an important example from the furniture manufacturer of the sort of technology that is converting manufacturing into an information based process. Machine vision does not mean recording or registering a raw image, as a camera would, but recognizing the actual objects in an image and assigning properties to those objects in order to understand what they mean. Vision in this sense makes every aspect of manufacturing—inventory, transport, tooling and assembly operations—much more efficient.

Machine Learning (ML) refers to a system capable of automatically integrating knowledge that it has learned from experience. This means that the system can improve itself from past experiences and therefore offers increased efficiency and effectiveness [4]. This involves tasks such as recognition, diagnosis, planning, robot control, prediction, etc. Machine learning is closely related to artificial intelligence and is based on proven technology that has had significant impact on both industry and science. There are numerous successful applications of machine learning, that serve to illustrate the diversity of possible applications, and the substantial benefits to be gained in virtually all sectors of the economy by applying ML to problems of process control, data analysis, and decision-making. The furniture manufacturer has ambitions to explore the possibilities within machine learning but has also experienced that the human aspects cannot be neglected. There will always be a necessity to make decisions that are based on irregularities or elements that are not included in the automated decision processes.

23.3 Automatic Data Collection for Creating Knowledge

Adaptive manufacturing is a term that covers many aspects of a manufacturing company. Definitions related to the term range from information handling and improving business, down to manufacturing operations and self-adjustable tools in machining. A basic element is to gather as much data automatically that can be useful for operations and improvements in manufacturing. This also means integrating information systems from operations at shop floor level up to strategic management and development.

Adaptive manufacturing normally implies that there are fewer people involved in manufacturing processes. In theory most of the operations could be executed and managed remotely by the mother plant, and where people from the different manufacturing satellites are mainly responsible for maintenance, security/control and to some extent problem solving when unexpected incidents occur. However, centralized control and decision systems have negative aspects that the furniture manufacturer has experienced. Automated processes effectively provides an enormous amount of data that could be very important in the knowledge creation and knowledge transfer processes. But this is mainly explicit knowledge and knowledge about customer trends and other facets of the business environment can easily be neglected or drowned in data overload. This knowledge is to a large extent about intangible aspects and is tacit. The furniture manufacturer has experienced how the reduced involvement of people in operations and process improvement has promoted disinterest and lack of motivation among employees. One of the results has been that the business improvement process has not been as fruitful as the top management would like. To increase motivation roles have been defined and production has become more team oriented with much more focus on personal experiences and tacit knowledge. These teams have not only tasks related to operations but even more important have defined responsibilities for knowledge creation and improvements.

References

1. Maier-Speredelozzi V, Koren Y, Hu SJ (2003) Convertibility measures for manufacturing systems. CIRP Ann Manufscturing Technol 52(1):367–370
2. SAP AG (2003) Manufacturing strategy: an adaptive perspective. SAP White Paper mySAP SCM. Available at http://www.sap.com/industries/automotive/pdf/BWPWP_Manufacturing_ Strategy.pdf. Accessed 10 April 2007
3. Røstad CC, Stokland Ø (2007) Adaptive manufacturing from enterprise to shop-floor level— sense, determine and execute. SINTEF Technology and Society, SINTEF rapport, STF50 A07035, Trondheim
4. Alberta Ingenuity Centre for Machine Learning (2007) Creating value for Alberta through excellence in machine learning. AICML strategic business plan 2007–2012

Chapter 24
Sustainable Manufacturing in SMEs—A Case from Sportswear Manufacturing (Case 4)

Abstract There is no doubt that sustainability is becoming more important in manufacturing and not only as a response to requirements from government regulations. We also see a growing number of companies having sustainability and social responsibility as the basic element of their core business model. In this case an example of a sportswear company is presented. We see some of the challenges facing the company when it comes to offshore manufacturing and how it has a particular focus on a work force that otherwise would have difficulties with entering the labor market.

There is no doubt but that sustainability is getting more important in manufacturing. This is not only as a response to government requirements but also a growing number of companies who have sustainability and social responsibility as the basic element of their business model. In this case study an example of one such a company within the sportswear industry is presented. We see challenges when it comes to offshore manufacturing decisions and how the company has a particular focus on people who would otherwise have difficulties entering the labor market.

24.1 The Strategic Context

The company has a focus on sustainability. Corporate Social responsibility (CSR) is the paraphrase that pervades the company not only in documents, homepage, speeches, etc., but also in real action and is also reflected in the manufacturing strategy. The strategic context of the company is illustrated in Fig. 24.1.

The small sportswear company was founded in 1998 and the founder still owns and manages the Company. The company has experienced a substantial and steady growth over the years, and has today gained a robust position in the market place. It has 80 employees and a turnover of 20 million Euro. The main product of the

A. Rolstadås et al., *Manufacturing Outsourcing*,
DOI: 10.1007/978-1-4471-2954-7_24, © Springer-Verlag London 2012

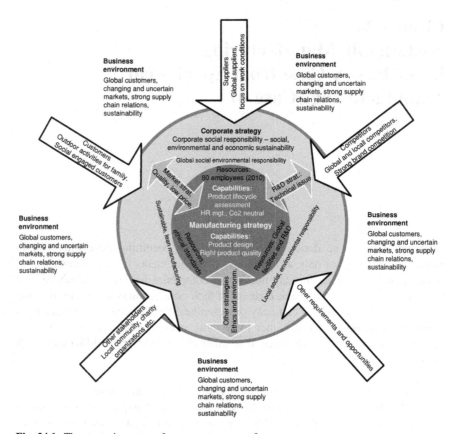

Fig. 24.1 The strategic context for sportswear manufacturer

company is the design and sale of functional clothes for outdoor activities. The clothes are produced by 14 factories in China, while administration, design, wholesale distribution, etc. is carried out from the headquarters.

From Fig. 24.1 we see that CSR is reflected in almost all elements of the strategic context of the company and in particular in their relations with the stakeholders and business environment. The company's mission is to contribute towards positive societal development by looking after each other and caring for customers' welfare. The general view of the relationship between the company and society is that the company is dependent on the well-functioning of the society that surrounds it. As citizens and participants in the labor market they need a society which functions. The managing director/owner states that he has always been socially-minded. He became particularly aware of the importance of social responsibility during his eight active years as a local politician. He served four years as an elected politician in the local council. He has also served on the Board for a Shelter for Battered (misused) Women and Children and therefore knows what the downside of life can be like. He believes that all persons must take personal responsibility and that human beings have choices. This belief forms a basis for all of the strategies in the company including the manufacturing strategy.

The market for sports outerwear clothes is difficult and is to some extent exposed to macroeconomic fluctuations. However, maybe even more important is branding and customers changing brand preferences. Our case company is working hard to gain and strengthen customers' brand loyalty. One of means they have used has been social media and a 'virtual society' related to the company's social responsibility. The company has 20,000 Facebook followers and customers who 'feel good' about supporting a civic minded company. In addition, the CEO personally invests a lot of time on social media to foster dialog around important issues ranging from ethical trading to climate and financial crises. In other words the company has created a basis for increased sales through their CSR branding.

The products range from underwear and swimsuits to outdoor clothing for cold climates. The company has a profile directed towards sports oriented families but that are not the most expensive in the market. The company designs and sells, among other things, children's outdoor apparel (clothes) where they are leading in designing 'safe' clothes at affordable prices.

Their concept is about making sports outwear that look good, have good functionality, but that don't costs too much. On their homepage they say announce lean manufacturing principles to reduce 'over-processing' and keep costs reasonable. Another important element of their manufacturing strategy is stable work force. This is a capability they have gained through their strategy of taking responsibility for hiring people that for most companies would seem inefficient, risky and costly.

24.2 Competitive and Stable Workforce Through Inclusion

The company goes to great lengths to be an inclusive employer. The company has signed the IW (Inclusive Work) agreement (see Chap. 2). One of their main approaches is recruiting people who would otherwise have difficulties entering the labor market. One in four employees in the company has had difficulties getting into the labor market. Some of them are convicted felons, former drug addicts and other people who have had trouble landing a job for one reason or another. According to their strategy at least 25% of their employees should be recruited from persons who for different reasons have difficulty with entering the labor market. At present nearly 30% of the employees fall into this category. The company collaborates in this respect with an organization, 'Way Back', that helps prisoners and former prisoners with their rehabilitation.

The owner who started the company more than 20 years ago, motivates employees through inclusion. Once a month, the company has a joint operations meeting where all employees participate in making company decisions. Because so many of the employees have received direct help from the company, in the form of a stable job, they express an eagerness to help. As such, the company is a large contributor to different humanitarian organizations, but also organize an annual event for women and children who live at several shelters in the country. According to the CEO such civic activities create positive employee morale and help employees feel connected to each other and to the broader community. The company's motto is: "We cannot change the world, but we can change the world for someone."

Behind this are two ideas: It is cheaper to retain existing workers than to acquire new ones, and engagement in the workplace ensures increased productivity. By building on passion and pride, companies do more than just business. They become beloved brands that live up to their promises.

Although recruitment among prisoners puts a demand on the company, training and culture wise, it has not cost the company anything. In return the company gets very loyal employees, resulting in very low attrition and absenteeism. Only three people have left the company since it was first established. The investments in ethical trading in China pays off by way of a stable and productive workforce. This is directly linked with competitiveness and the fact that the company is a profitable business (and growing) and has developed a strong brand in relation to both national and international brands in the field of outdoor activity garments.

What the company has experienced through their strategy of recruiting people that have dropped out of normal work is that among these people and people around them, and even from local authorities, there are a lot of engagement and positivism that the company will benefit from. In return, the company's employees must meet high performance goals. If companies' strong sales results are any indication, it's clear that when given a chance, socially challenged employee groups perform at their best and live up to high expectations. A stable workforce lets the company concentrate fully on winning share in a competitive market. This is an example of how corporate social responsibility and sustainable manufacturing goes hand in hand with business operations. The company calls such an approach "value for all".

24.3 Manufacturing Units in China Included in the Sustainability Strategy

The sportswear company is European but to be competitive the company has seen it necessary to manufacture in regions that could do it to a lower cost. They have for example 14 factories in China manufacturing for them. However, the company will not compromise on ethical standards and other aspects of their CSR policy. Suppliers and partners are followed closely to assure that they apply to the standards.

The companies' codes of conduct have been translated into Chinese and have been placed in a public area within each of the factories so that the Chinese workers can read them. In cooperation with an ethical trading association they have developed instruments for inspecting the Chinese factories once to twice a year.

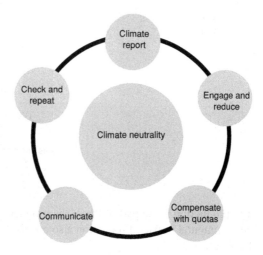

Fig. 24.2 The main elements of climate neutrality in the sportswear company

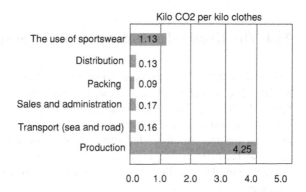

Fig. 24.3 CO_2 footprint (kilo CO_2 per kilo clothes)

This association is together with trade unions also represented in these inspections. The inspections are sometimes announced in advance, and sometimes not. The inspections/interest is believed to have improved labor standards in the factories, but they admit that is much left to improve.

Climate neutrality is an important part of the strategy of the company and is also a premise for the whole supply chain. In Fig. 24.2 the main elements of climate neutrality is presented. As we see there is an implicit aspect of improvement in this framework. However, in this model we also see that there are challenges in engaging and involving subsidiaries, suppliers and (distant stakeholders) in this improvement process. Life cycle analysis performed by external resources is a basis for this model. Life cycle analysis and the reporting is one challenge, but the fact oriented and explicit character of the reporting makes it less challenging than the more contextual aspects of, involvement, motivation, communication, etc. needed in the improvement process.

Figures 24.3 and 24.4 are illustrations and examples from the climate report of the company. We see that the CO_2 footprint also covers the use of the sportswear.

Fig. 24.4 Distribution of
CO$_2$ footprint

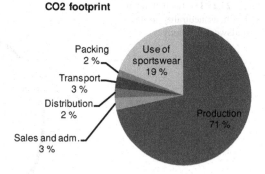

The production is the part of the of the product lifecycle that has the most negative environmental impact. Production also covers the manufacturing in the 14 Chinese factories.

24.4 The Knowledge Dimension of the Sustainability Strategy

CSR and sustainability are meant to be the basic premises for manufacturing also for subsidiaries and suppliers in other regions. However the company admits that this is challenging. One of the challenges is related to the fact that the sportswear company is rather small compared to other customers. Hence their supply chain power is limited. Another challenge is how to motivate for a focus on sustainability in factories and regions very different from what we have in the European mother plant. This challenge is also reflected in the knowledge dimension of the manufacturing strategy.

First of all CSR and focus on sustainable manufacturing requires a very high degree of knowledge creation and knowledge transfer between people and between the stakeholders. There are new and different elements that are added to the strategy and knowledge related that is to a large extent tacit and contextual. The way the sportswear company has approached these challenges could be grouped into four:

- A global CSR strategy that also is reflected in the manufacturing strategy throughout the supply chain
- Partner selection where one of the criteria is a common understanding of basic manufacturing principles based on sustainability
- Visits and personal contacts with key persons in factories and other partners worldwide. This reduces contextual barriers in knowledge transfer especially of tacit knowledge
- Audits and different types of reporting to bring as much facts as possible to the table regarding sustainability.

Even if the company has a global CSR strategy and focusing on sustainability in their manufacturing the company has experienced that some compromises sometimes have to be made. This concerns in particular timing in the implementation of principles, etc. The founder and CEO of the company is very visionary and a facilitator of the strategy. This is one of the reasons for the company to put extraordinary energy into implementing a complex strategy.

About the Authors

As complexity in manufacturing increases we need to develop new approaches, new knowledge, and new ideas for how manufacturing organizations can be transformed back into 'competitive weapons' for industrialized countries. Knowledge is a broad field of study, and we as engineers, managers and economists have to find ways to create disciplines and professions that deal with the challenges faced by manufacturing organizations. The authors of this book represents a multidisciplinary view of the issues faced by manufacturing organizations and between them have an extensive research and applied knowledge of the development of manufacturing.

Bjørnar Henriksen is a Senior Research Scientist at SINTEF Technology and Society, Norway. He is also a research fellow at the Faculty of Engineering Science and at the Norwegian University of Science and Technology, from where he also obtained his PhD within the field of manufacturing strategy. Henriksen has a broad experience in consulting and applied research from large international consulting firms. Process improvement, reengineering and product development have been important topics in these projects within manufacturing industries. Today Henriksen is project manager for several large research and development projects funded by the Norwegian Research Council. Enabling technologies, performance and knowledge management are key elements in these projects. Henriksen is the author of three books and has also been a contributor to research in manufacturing through international conferences and journals.

Asbjørn Rolstadås is the Vice Dean for research at the Faculty of Engineering Science and Technology and a Professor of Production and Quality Engineering at the Norwegian University of Science and Technology. Professor Rolstadås has over 40 years of experience in education, research and consulting. His research covers topics like numerical control of machine tools, computer-aided manufacturing systems, productivity measurement and development, computer-aided production planning and control systems and project management methods and systems. He is the author of 13 books and over 280 papers. He is past president of the Norwegian Academy of Technical Sciences and member of The Royal Norwegian Society of Sciences as well as the Royal Swedish Academy of

A. Rolstadås et al., *Manufacturing Outsourcing*,
DOI: 10.1007/978-1-4471-2954-7, © Springer-Verlag London 2012

Engineering Sciences. Professor Rolstadås is the founding editor of the International Journal of Production Planning and Control, and is past president of The International Federation for Information Processing.

David O'Sullivan is a Lecturer, Research Director and Director of Quality at the National University of Ireland, Galway. In addition to academic roles, he facilitates University management with strategic and operational planning, quality assurance and enhancement, performance measurement and innovation management. David also advises industry on how to improve their innovation and project management processes and recent projects have included leading multinational organizations such as IBM, Ingersoll-Rand, Fujisawa, Hewlett-Packard and Boston Scientific. He has also worked with a large number of small to medium sized industries. His research interests include applied innovation that he describes as the marriage between innovation management, knowledge management and project management. His current research interests are in distributed innovation management across extended enterprises. David has over 100 publications including books—Applying Innovation (Sage); Manufacturing Systems Redesign (Prentice-Hall); Reengineering the Enterprise (Chapman & Hall) and; The Handbook of IS Management (Auerbach).

Index

A

'talent-driven innovation, 50
spiral approach, 140
A3 report, 165
Absorptive capacity, 150
Acquisition cost, 100
Adaptive factory, 213
Adaptive manufacturing, 7
Agency theory, 120
Analytical knowledge base, 151
Arbitrary specialized, 149
Automated processes, 215
Automatic data collection, 215
Automatically adjust, 214
Automotive industry, 116

B

BEEM, 161, 180
Behavioral, 111
Benchmarking, 79
Branded goods, 211
BRIC countries, 44
Business environments, 16
Business systems, 199

C

Capabilities, 6, 24, 159
Capacity, 88
Categories of quality, 15
Center of gravity, 176
Climate neutrality, 221
Climate report, 221
Clustered companies, 171

CO_2, 60
Codes of conduct, 220
Codifying the tacit knowledge, 161
Coding/decoding, 159
Collaborators, 200
Communicating the vision, 133
Communicators, 199
Community, 138
Company functional level, 81
Competence, 147
Competitive forces, 20
Competitive weakness, 49
Competitiveness, 50
Concurrent engineering, 198
Contexts, 153
Context-specific, 160
Continuous improvement, 122
Continuous learning, 202
Continuous technology devel, 3
Continuously improve energy efficiency, 188
Contractor, 97
Co-operators, 200
Coordinating innovation processes, 134
Coordination challenges, 105
Coordination mechanisms, 119
Coordinators, 200
Corporate Social Responsibility, 181, 217
Cost management, 30
Counted, 72
Craft manufacturing, 7, 91
Craftsman, 207
Creating value, 91
Criteria-based, 112
Crystallized intelligence, 148
CSR strategy, 223

A. Rolstadås et al., *Manufacturing Outsourcing*,
DOI: 10.1007/978-1-4471-2954-7, © Springer-Verlag London 2012

C (*cont.*)
Customer requirements, 35
Customization, 9

D
Data overload, 72
Decision process, 99
Definitions of innovation, 57
Developing economies, 46
Digital business, 12
Distance, 170
Distributed structure, 105
Drivers for change in manufacturing, 39

E
Easy to measure, 82
Economic advantages of outsourcing, 97
Economic situation, 45
Educational level, 54
Effectiveness, 70
Emergent approach, 132
Energy consumption, 182
Energy intensive industries, 187
Environmental impacts, 192
Environmental indicators, 60
European States, 56
Evolving markets, 47
Evolving paradigms, 27
Explicit knowledge, 8
Extended enterprises, 12

F
Facilities, 88
Fact based knowledge transfer, 202
Factory focus, 64
Feedback system, 210
Financial crisis, 45
Five-axis milling machines, 214
Foresight, 5
Formal reporting structures and systems, 206
Formal/structural, 121
Frameworks for strategic performance
    measurement, 74
Full scale implementations, 207
Furniture manufacturing, 211
Future manufacturing, 4

G
GDP growth, 61
General world knowledge, 148

Generic strategies, 22
Geographical dimensions, 100
Geographical footprint, 101
Global agreements, 61
Global units, 167
Globalization, 11, 163

H
Heavy energy consumer, 183
Hierarchy of objectives and goals, 113
High end, 211
Hofstede, 125
Human factors, 92

I
ICT, 170
Ideation, 161
Implementing strategies, 183
Improvements, 36
Improving quality, 137
IMS2020 roadmap, 65
Incremental innovations, 141
Indicators for lean manufacturing, 77
Industrial Change, 57
Industrial Renewal, 53
Informal communities, 150
Informal, 122
Information and communication
    technology, 44
Information, 147
Infrastructure for knowledge transfer, 209
Infrastructure, 87
Innovation funnel, 138
Innovation process, 129
Innovation, 12
Innovative products, 57
Institutional, 111
Intangible aspects, 72
Internationalization, 102
Interplay of decisions, 92

K
KAT Roadmaps, 67
Knowledge base, 144
Knowledge base, 151
Knowledge creation spiral, 38, 192
Knowledge creation, 9, 37, 134, 158
Knowledge dimension, 8, 174
Knowledge production, 158
Knowledge society, 56
Knowledge transfer, 36, 179

L

Lean manufacturing, 7, 29
Lean, 177
Learning Economy, 58
Learning process, 162
Leisure boat manufacturing, 205
Life Cycle Analysis, 32
Life Cycle Assessment, 186
Life cycle focus, 64
Life-long learning, 58
Linear model, 143
Location decision, 109
Location issues, 102, 168
Location, 100
Loose federation models', 110

M

Machine Learning, 215
Macroeconomic fluctuations, 219
Make or Buy, 95
Manufacturing context, 173
Manufacturing paradigm, 90, 175
Manufacturing strategy, 20
Manufacturing structure, 93
Market-based view, 20
Mass manufacturing, 7
Matrix-methods, 190
MDP, 99
Measure performance, 72
Measuring, 69
Medium sized enterprises, 205
Mobilize tacit knowledge, 160
Modularization, 198
Module-based manufacturing, 14
Monopoly position, 208
Mother plant, 105

N

Neoclassical, 111
New economy, 13

O

OECD, 63
OEM, 200
Off-shoring, 98
Old Economy, 13
Open Innovation, 153
Operating policies, 104
Operational capability, 24

Operative paradigm, 28
Organizational boundaries, 37
Organizational configuration, 104
Organizational culture, 121
Organizational: and Cross Organizational
    Learning, 190
Oslo Manual', 130
Outsourcing units, 167
Outsourcing, 96

P

Paradigms, 6, 177
Parent companies', 107
Partner selection, 103
Passion and pride, 220
PDCA, 38, 140
Performance in supply chains, 82
Performance measurement framework, 77
Performance measurement, 70
PMO, 91
Portfolio of innovation processes, 169
Portfolio' of actions, 139
Process technology, 88
Process, 131
Productivity improvements, 49
Productivity, 15
Project execution, 124
Project model, 135
Project oriented teams, 209
Project, 123

Q

Quality Function Deployment, 76
Quality improvement, 202
Quality, 15
Quantifiable, 72

R

R&D internationalization, 136
R&D models, 139
Radical innovations, 141
Rrankings, 50
Rationalism, 146
Repertoire of knowledge, 39
Research programs, 61
Research roadmaps, 63
Resource-based view, 20
Resources, 23
Rhetoric focused, 55

R (*cont.*)
Robots, 212
Rocess innovation, 132

S
Satellite structure, 105
Self-assessment, 79
Shop floor, 135
Simplified LCA, 189
Skills, 25
SMED, 78
SMEs, 217
Social responsibility, 4, 31
Socially-minded, 218
Spatial distance, 171
Specific strategies, 22
Sportswear company, 217
Stable workforce through inclusion, 219
Stakeholders, 74
Strategic context, 19, 23
Strategic criteria, 113
Strategic decisions, 177
Strategic performance approach, 74
Strategic scope, 22
Strategic strength, 22
Structural prerequisites, 89
Structure, 87
Structured approach, 112
Supply chain focus, 64
Supply chain integration, 197
Supply chain relations, 199
Supply chain, 82
Sustainability, 15, 60
Sustainable manufacturing, 7, 30, 116, 177
Synthetic- and analytical knowledge base, 151

T
Tacit knowledge, 149
Tacit–explicit dimension, 9
Tactical and operational decisions, 32
Tactical processes, 107
Talented workers, 50
Team leader meetings, 210
Teams, 138
Technical knowledge, 73
Technological adoption, 47
Technology Outlook, 44
The Global Competitiveness Report, 49
The Lisbon Agenda, 55
Tier 1 supplier's, 201
TOPP, 79
Total Quality Management, 131
Traditional manufacturing, 145
Training methods, 68
Transactions, 100

U
Ubiquitous learning, 198
Uncertain Future, 5

V
Value chain, 150
Vertical integration, 88, 201
Views on knowledge, 146